CURING MS

ALSO BY THE AUTHOR

FICTION

The Children's Ward

NONFICTION

Neurology for the House Officer

CURING MS

How Science Is Solving the Mysteries of Multiple Sclerosis

Updated with the Latest Treatment Information

HOWARD L. WEINER, M.D.

THREE RIVERS PRESS • NEW YORK

Published in the United States by Three Rivers Press, an imprint of the
Crown Publishing Group, a division of Random House, Inc., New York.
www.crownpublishing.com

THREE RIVERS PRESS and the Tugboat design are registered trademarks of Random House, Inc.

Originally published in hardcover in the United States in slightly different form by
Crown Publishers, an imprint of the Crown Publishing Group,
a division of Random House, Inc., New York, in 2004.

Library of Congress Cataloging-in-Publication Data
Weiner, Howard L.
Curing MS : how science is solving the mysteries of multiple sclerosis /
Howard L. Weiner.—1st ed.
1. Multiple sclerosis—Treatment. 2. Multiple sclerosis—Psychological aspects.
[DNLM: 1. Multiple Sclerosis—therapy—Personal Narratives.
2. Multiple Sclerosis—psychology—Personal Narratives. 3. Neurology—
Personal Narratives. 4. Research—Personal Narratives.
WL 360 W423c 2004] I. Title.
RC377.W446 2004
616.8'34—dc22 2003023408

ISBN-13: 978-0-307-23604-3
ISBN-10: 0-307-23604-8

Printed in the United States of America

DESIGN BY BARBARA STURMAN

10 9 8 7 6 5 4 3 2

First Paperback Edition

For all those struggling to cure MS and

for all those who must struggle until the cure is found.

Contents

CURING
MS

Introduction

I RECENTLY READ a book about climbers on Mt. Everest, and then I flew over Everest in a two-engine propeller plane. Although I never plan to climb Everest myself, for a period of time as I read the book I was on the mountain, my heart pounding as I vicariously moved from Base Camp to Camp One and Camp Two on the way to the summit. For me, trying to cure multiple sclerosis has been as exciting and even more frustrating than I imagine climbing Everest to be. The tempo and time frame are different. There are more unknowns. The stakes are higher. No one's been there before. This book is the story of those on the mountain trying to solve the mystery of MS, trying to find the elusive cure. It is also my own story, the story of my time on the mountain and how I've fared, and the story of the many MS patients I have cared for over the years. While on the mountain we will not only learn about multiple sclerosis but learn how science is practiced in today's age and what it takes to solve a medical mystery and cure a disease. I will serve as guide.

Multiple sclerosis is the number one cause of paralysis in young people. It is the polio of our age. I first encountered the disease as a medical student and I can remember well the case of the first MS patient presented to me in medical school. A twenty-six-year-old woman had suddenly gone blind in one eye and then almost as quickly recovered. Three years later, while having a drink with one of my high school buddies, Normie, I learned that six months earlier he too had gone blind in one eye. Five years later, during my neurology residency at Harvard, I faced the disease close-up when I treated a patient named John Saccone. Taking care of John Saccone affected me. He was my age. He had young children, as did I. I realized that I could easily have been him. Following that experience, I embarked on a lifelong study of MS in an attempt to find the cure. I put myself on the mountain with the others, an adventure that has lasted for more than three decades and still continues.

A close friend of mine is the writer Leslie Epstein, who heads the creative writing program at Boston University. Over the years we have spent countless hours talking philosophy, sports, writing, and medicine. I cherish his insights and wonderful wit. He helped me publish a novel I wrote a number of years ago, and for many years he has encouraged me to write about multiple sclerosis, to tell the story. I have always resisted, not because I didn't want to write about MS but because I didn't feel we knew enough about the disease for me to write about it in a meaningful way. The story was not ready to be told. That has changed. For the first time we have treatments that are beginning to help MS patients, and we have a clearer picture of the disease and what causes it.

The story of multiple sclerosis is actually many stories. Foremost, it is the story of those that suffer from the disease, of their courage in fighting the disease, and of the impact of the disease on their families. It is the story of those who have successfully fought the disease and those who haven't. It is the story of how difficult it is for a person to deal with an unpredictable chronic disease. MS patients have fears and many misconceptions about the illness; they need hope. Furthermore, experiments must be performed on MS patients if the disease is ever to be cured; countless patients have volunteered over the years. Thus, the story of multiple sclerosis is also about the moral issues that must be confronted when the physician and patient form the only team that can test a cure for the disease. How does a doctor support patients and provide hope without false optimism in the midst of a clinical trial or research experiment?

The story of MS is the story of the scientists and physicians who are struggling to understand the disease, and their personal hopes and frustrations. Scientists and physicians are fallible, given to jealousy, and can be territorial, qualities that often impede the process of discovery. Nonetheless, in science human frailties are counterbalanced by the ultimate truth that's waiting to be discovered, and if a scientist's frailties leads the search in the wrong direction, the biological truth ultimately keeps the search in line. As a television commercial comparing margarine and butter once proclaimed, "You can't fool Mother Nature." The disease process is both cruel and mysterious, and it frustrates all of us who study it, but it follows biological principles, even if it is nature gone astray. Mother Nature will impartially answer any question posed, but the questioner must be able to pose it correctly and understand the answer. Thus, even though most scientists and physicians have their biases, Mother Nature keeps them honest. We will see how scientific questions are posed, how the scientific method works and doesn't work, and the problems associated with using it.

The story of MS is the story of competition between scientists. Who's first? Who will win a Nobel Prize? It is part of human nature to compete and to give prizes: who is the prettiest, which film or song is the best, and who can run, swim, jump, or hit a ball better than anyone else. Prizes are given in science as well, and some scientists make a bigger contribution than others. But when I think about competition between scientists and the scientific process, I often think about Alfred Wegener, a German experimental meteorologist who postulated at the beginning of the 20th century that the continents of the earth had once been attached to each other in a single supercontinent called Pangaea and then separated. Many a schoolchild has looked at the globe and noticed immediately that the west coast of Africa and the east coast of South America fit together like two pieces of a puzzle. But Wegener was ridiculed for his idea of floating continents. How in the world could continents possibly move? In 1930, while on an expedition to Greenland, he was caught in a storm, and it became clear that he would not survive. How did he feel about his ridiculed theory of floating continents and if he was right that he would not live to see its ultimate proof? Before he died, he wrote this in his journal: "It doesn't matter what happens to individuals, science is a social process that happens on a time scale longer than a single human life. If I die, someone takes my place. If you die, someone takes your place. What is important is to get it done." We now know that Wegener was right; the continents were once all together. In the 1960s (thirty years after his death) scientists discovered an explanation for Wegener's theory. Deep on the ocean floor, the earth is moving and shifting according to a geological theory called plate tectonics, a theory that explains earthquakes, volcanoes, the formation of mountain ranges, and why the continents separated.

Despite all of its competitiveness, one of the beauties of science is that it is a collective effort. A corollary of this is that whatever is discovered or hypothesized by one scientist must be confirmed and validated by other scientists. All scientists are dependent on each other, all of their work is interwoven, and no one is beyond making a mistake. Isaac Newton wrote that he could see as far as he could because he stood on the shoulders of giants. Einstein said that his own work would have been impossible without Newton's discoveries, and his big blunder was postulating that the universe was static until Hubble showed it was expanding. The story of solving the mystery of multiple sclerosis, like the story of all science, is thus the tale of a collective effort in which each person builds on what has been discovered by someone else. There have been a lot of people on the mountain.

The story of MS is the story of the immune system, the body's defense against disease, and it is the story of viruses and how they interact with the immune system. When I embarked on the study of MS in 1972, immunology and virology were two of the most fascinating basic science disciplines, and research in these fields has produced seven Nobel Prizes in the past twenty-five years. We will explore the compelling account of the workings of the body's immune system and how it has yielded its secrets to scientists over the past quarter century. When I began studying MS, the immune system was understood at a relatively simple level: it was composed of antibodies and white blood cells, barely distinguishable under the microscope. These once indistinguishable cells are now recognized to be a complex interacting network, with more classifications and subdivisions than ever imagined. The immune system touches all parts of medicine, including organ transplantation and the treatment of cancer, atherosclerosis, and AIDS. Other diseases that, like MS, are classified as autoimmune disorders include juvenile diabetes and rheumatoid arthritis. In telling the story of MS, we will learn about the immune system and viruses and the search for an answer to how viruses are related to the disease.

The story of MS is also the story of the impact of technology and business on medicine. So much of discovery is driven by technology, and one of the biggest breakthroughs in our understanding of MS was the development of magnetic resonance imaging (MRI) to image the brain. Imagine trying to study heart disease without a cardiogram or tuberculosis without a chest X-ray. Another breakthrough was the development of recombinant DNA techniques, which gave us the ability to genetically engineer proteins. The first drugs approved for MS by the Food and Drug Administration (FDA) were genetically engineered interferons. We will examine how advances in technology are crucial and may be the driving force in shaping our fight against the disease.

Then there is Wall Street. Developing a new drug for any disease requires hundreds of millions of dollars. That level of funding can come either from the government or from the private sector, and in the United States, the private sector is largely responsible for drug development. We will examine how the first drugs for MS were discovered, tested, approved, and marketed. The drugs are now priced at over $10,000 per year per patient. One of the discoveries in my own laboratory led to the creation of a biotechnology company, and I was suddenly forced to learn things they never taught in medical school.

The story of MS is the story of the lay press and how it reports science and medical research. In medical coverage, the media most like to tell about breakthroughs. The media have enormous power in shaping public opinion and in focusing the public's attention. The leading medical journals compete furiously to get press coverage and send advance copies of their issues to the mass media. News of many medical breakthroughs is first announced on television. Anyone who publishes scientific articles in one of the leading medical or scientific journals cannot escape the media. I have been on the national news several times, and for a scientist this is a true double-edged sword. It is difficult for patients as well, who struggle to understand the significance or robustness of the latest breakthrough from sound bites on the evening news. We will examine medical reporting by the lay press and how to interpret it.

The story of MS, like the story of science itself, is the story of many failures. Even though the past thirty years have witnessed breakthroughs in our understanding and treatment of MS, the road has not been easy. The term *multiple sclerosis,* literally translated, means "many scars." In a sense the story of MS is the history of the many scars that have accumulated over the years as scientists have tried to solve the mystery of what causes the disease. I myself have accumulated more than my share of scars over the past thirty years.

Those years have been spent in laboratories and hospitals associated with the Harvard Medical School in Boston. I often think of the scars when I walk across the Harvard Medical School quadrangle and stop to read a set of words etched onto the marble facade of one of the buildings. I discovered the words in 1976, when I was studying how viruses infect the brain. I have always liked them, and as the years have passed I understand them better, a bit like Mark Twain, who discovered how much his parents had learned during the time he aged from eighteen to twenty-three.

The words are not easy to read. Taken from the Hippocratic oath, they were inscribed in 1904 to inaugurate the Huntington Laboratories for research in pathology, and they show the wear of a century of New England winters. The words appear as one phrase under the other, almost like a poem: "Life is short, and the art long / The occasion instant / Experiment perilous / Decision difficult." Every time I read these words I realize how difficult it is for one person in one lifetime to understand biological processes that evolved over millions of years.

Although we've made so much progress, why is there still no cure for MS? Why has it taken such a long time? This is the number one question I am always asked by patients and their family members. First, MS is a chronic disease, and it usually takes fifteen to twenty years to run its course. Thus the testing of new treatments takes time; clinical experiments are long ones. Second, and even more important, MS is a very complex disease whose cure depends on understanding a large number of variables, some of which still remain unknown, such as what causes the different subtypes of the disease, what happens when a patient enters the progressive stage of the disease, and why some people respond to treatment while others don't.

Variables. Medical science is the investigation of biological variables created and modified by evolution. The scientific process asks questions in order to identify new variables and then tries to understand how the variables interact with one another, like numbers in an equation. A mathematical equation such as $2 + X = 4$ is easy to solve; there is only one unknown. An equation in biology, A + B + C + unknown variables = X, is far more difficult. The multiple sclerosis equation looks like this: Multiple sclerosis = virus + immune system + genetics + environment + gender + geography + age + unknown variables. In learning the story of MS, we will investigate each of the variables in that equation and try to understand how they fit into the ultimate solution. We will also explore which variables are still a mystery.

In the novels I have written, there is a main protagonist, usually a doctor. In this book, the main character is the disease itself: how it has affected patients, how it has frustrated scientists, how it has affected me. MS is a disease with its own personality. Once we understand what crucial factors during the course of evolution conspired to bring it into being, we will then know what is needed to put it to rest.

THE TITLE I CHOSE for this book, *Curing MS: How Science Is Solving the Mysteries of Multiple Sclerosis,* implies that finding the cure for MS is a work in progress. Thus, *Curing MS* is intended as a chronicle of the ongoing story of how science is moving closer to that goal. I tell the story of MS in chronological fashion, beginning with my first experiences with the disease in 1971. Fortunately, the field is moving quickly, and it is my intention to add important milestones and breakthroughs as they occur. This paperback edition adds a chapter that tells the story of the newest drug that was approved by the FDA for MS and of drugs that are waiting in the wings. The last chapter describes the latest progress in "going for the cure" and the new strategies being employed to get there; in the glossary, I have provided a link to yearly updates on MS from our MS center. One day, I hope the title of the book will be changed to *Curing MS: How Science Solved the Mysteries of Multiple Sclerosis.*

1

THE MONSTER

OVER THE YEARS, I have become friends with the Israeli writer Aharon Appelfeld. Aharon was born in a town called Czernowitz in the former Hapsburg border province of Bukovina. As a child he was deported to a concentration camp in Transnistria. He survived and settled in Israel, where he has lived for the past fifty years and where he writes about the Holocaust in a subtle yet powerful voice. We have shared many cups of coffee together in his apartment outside of Jerusalem, and I have shown him the centrifuges, freezers, and animals in my laboratory in Boston. In his life and work Aharon has had to confront the human spirit gone astray; in my work I have had to confront biology gone astray. Two different holocausts.

"The disease you are studying is a monster," Aharon said to me late one afternoon as we walked in the hills surrounding Jerusalem. "You cut it here, it grows there. You think it is down, but it rises again. You believe you understand why it behaves the way it does, but you are wrong. That's what you must do, Howard, tell the story of the monster."

Aharon was right. Multiple sclerosis is a monster. The monster has finally been wounded, but not yet killed.

I confront the monster every time I see a patient with multiple sclerosis, an experience that has been part of my life for over thirty years. There are over four thousand patients seen in our MS center every year, and I have patients I have followed for close to twenty-five years. I no longer see general neurology patients. Although I have worked hard in the laboratory and now run my own large lab, I never just wanted to work in the lab; I always wanted to confront the disease face-to-face. What one discovers in mice may ultimately be crucial to understanding the disease, but it is always one step removed. I have learned from my patients by studying them directly, they are a vital part of solving the riddle of MS. Thus I confront the monster not in the abstract but in the individuals I encounter every day.

I have had countless experiences with my MS patients over the past thirty years. Some have done extremely well; others have fared much worse. There are patients who share my passion for golf and not only trade golf stories but describe their illness in terms of how the disease has affected their game or ask whether they will ever be able to play golf again. There is the woman I had not seen since I treated her for an MS attack she experienced twenty-two years earlier, after the birth of her daughter; she brought her daughter to me because her daughter had had her first MS attack. There are patients I see year after year who have done so well with their illness that I spend more time talking about our families than their MS. Then there are the doctors with MS—family practitioners, surgeons, and even neurologists, with whom I speak in sophisticated medical terms but who transform from doctor to patient in the examining room. There are fighter pilots and movie stars. There are patients in wheelchairs telling me not to give up hope in my research, telling me that they don't hold me personally responsible for not curing them. There are those who tell me we are not working fast enough. There are patients who have traveled across the country or across the ocean to see me. There are patients who participate in clinical trials and donate blood. There are elderly parents bringing in a middle-aged child, and there is the occasional child or young teenager with MS. There are patients I have stopped from going into a wheelchair or brought out of a wheelchair, and there are patients who have ended up in a wheelchair despite all I did. There are patients who come in with long lists of questions and the latest newspaper article reporting on MS breakthroughs. There are those asking advice about quack cures, and there are those with sophisticated questions. It is emotionally difficult for patients to hold their hopes in check, and it is just as difficult for myself and my fellow researchers who have to play the part of instilling hope, only to see that hope dashed when a treatment disappoints.

Over the years there have been certain moments when the monster has come into painfully clear focus. One of these moments occurred in the spring of 1999 when I stopped in Denver to visit my mother on the way back from an MS fund-raiser in Los Angeles. My mother, who was almost eighty but in wonderful health, picked me up at the airport, and instead of driving south to her apartment we headed north to an assisted living facility called Mary Crest. My childhood friend Norm Wedgle had just entered the facility because of his MS, a disease that began thirty years earlier, though most MS patients do not require such a facility.

Norm loved life. In high school he was the jokester and I was the serious one; because of this we balanced each other and found both solace and joy in confronting the early stages of our lives together. I went east for school, and he took over his father's soap manufacturing business. Norm liked business. He liked to put together deals, and he was a salesman, always pushing something, always telling a joke, not easily discouraged.

Norm had his first MS attack in 1968, as I was finishing medical school. It was optic neuritis, blindness in one eye. The blindness lasted for six weeks and then resolved. For many years Norm did well with his illness, but as time passed the disease began to take its toll. I saw him from time to time when I visited Denver, often meeting him at my parents' delicatessen. In 1999 I hadn't seen him for almost five years. When my mother told me he had entered an assisted living facility, I suggested that we visit him. His sister gave us the address, and my mother said, "Let's surprise him."

I had spoken to Norm occasionally over the years, and even though we were both now in our mid-fifties, to me he was Normie and to him I was Howie. In high school Normie had given us nicknames patterned after our body shape—I was Big Skinny and he was Little Fats.

As we approached the assisted living facility, I was afraid that he might not be there and asked my mother whether we had made a mistake by not calling or arranging the visit in advance.

"Don't worry. He'll be there, and he'll be so surprised to see you," she said. She was right.

"Room 109," the receptionist said. "Norm is in his room watching TV."

The door was ajar, and I knocked on it and pushed it open. When Normie asked who it was, I said, "I've come to see Little Fats." Without missing a beat Normie yelled out, "Big Skinny!" He turned around in his chair, and we gave each other a hug.

I sat down next to him, and we held each other's hand for a moment. Normie was dressed in a T-shirt and athletic warm-up pants. He had a full beard. I looked around the room. Next to him was a motorized wheelchair. When we were teenagers, Normie had had a motor scooter, and I spent hours on the back of it, chasing girls and stopping for banana splits at the Dairy Queen. Now Normie was no longer able to walk, but he could stand up and transfer from his chair to the wheelchair. This seemingly small act gave him critical independence. It meant he could leave his room on his own and visit with others in the assisted living facility. It meant he could bathe on his own by transferring from his wheelchair to a shower chair.

Normie still had his wits about him, and during our conversation he was quick to crack a joke. He had always been a jokester. When my mother asked him how old he was, he smiled and said, "Lola, there are only three ages. You're young, you're middle-aged, and you're looking good!" We all laughed. Normie and I were the same age, fifty-five, and he remembered that my birthday was on Christmas.

As Normie spoke I could hear the cadences of MS: his speech came out in uneven bursts. The medical term is *scanning speech,* and the phenomenon is caused by damage to the brain stem, where motor fibers that control the speech muscles are located. Normie showed signs of other brain stem problems as well, as his eye movements were jerky and irregular. It was disorienting for me. I felt drawn to him as my childhood friend but also found myself reacting to him as a physician, trying to assess his condition, something that required emotional distance. I noticed several family pictures on the dresser and asked if he minded if we looked at them. But when I handed him a picture and put it in his right hand, he dropped it. The right hand was not weak but it was clumsy, and he told me he had to use his left hand for writing and eating. But Normie was keeping up his fighting spirit. After he explained to me the problem with his right hand, he shrugged and said, "It's the worst thing that should ever happen to me." Normie coped by always looking on the positive side, something that is not easy for many MS patients. Nonetheless, each patient ultimately finds his or her own way to cope. Normie was the first resident of the Mary Crest home, and he was proud of it. He was currently lobbying to get an exercise room for the facility and told me about ideas he had for fund-raising.

The first picture we looked at was of the two daughters he had with his first wife, Sandy. I knew Sandy; we were all classmates together in high school. One of his daughters was married, and Normie was a grandfather. "Howard," my mother said, "see, Normie's a grandfather. It's time for you to become one too."

Normie's first marriage had broken up, and he then married another woman who also was named Sandy. "I called her Sandy number two," Normie said, "but we broke up because of the MS. She just couldn't handle it."

"Is there a Sandy number three?" I asked.

"Sandy number three's name is Linda. She's my girlfriend here in the home," Normie said. "Later we'll go see her."

Of all the pictures we looked at, the one that struck me most was a picture of Normie as a young man, a high school student. That is how I remember him: handsome, chiseled features, a full head of hair. His facial features had changed, as they do for everyone as they age, but because of his illness it was more striking.

Normie had only been in a wheelchair for the past two years, but in the previous five years his condition had been slowly worsening, following a typical pattern for MS. His disease began with an attack from which he fully recovered—blindness in one eye when he was twenty-five years old. Over the next twenty years, in the relapsing-remitting stage of the disease, he suffered intermittent attacks, weakness in the legs, loss of balance, and problems with sensation. These attacks incurred relatively mild disabilities, not affecting his life in a major way. But then, as is often the pattern, his disease became progressive and he steadily worsened without attacks, with the primary effect being on his walking. A major question in multiple sclerosis is why the disease changes this way, from the relapsing-remitting stage to the much more debilitating progressive form.

Normie was on a fat-free diet and he had several diet books in his room. Normie was one of the many people who have tried various diets for MS, though there was no evidence that special diets helped MS. He was also taking one of the new injectable MS drugs. Unfortunately, I did not know how much any of these things would help him. He would have had to be taking the drug since the first sign of his MS, almost thirty years ago, for it to have had a chance to have a major impact on his disease. The progressive form of MS is the most difficult to treat.

Oh, how I wished there were something I could do to help him. I thought of other drugs he might try, of new drugs we would be testing, and then asked him the name of his physician. "Bowling," he said. When I asked him what Bowling's first name was, he said, "Doctor." I smiled. Few patients call me by my first name. I knew Bowling; I had met him at an MS conference, and he wrote a book on alternative medicine in MS. I made a note to call him and discuss Normie's case, though I was doubtful that there was anything I could do that would have a major impact on Normie's illness.

"Come, I'll show you around," he said as he transferred into his wheelchair and gave us a tour of the home. He introduced me to people as his friend Dr. Weiner, "the number one researcher on MS in the world." I was a bit embarrassed, to say the least, until he turned to me and said, "Big Skinny, you promised me thirty years ago you would cure MS, and you still haven't done it." My embarrassment turned to frustration, and I tried not to show that either. I thought of all the talks over thirty years in which I had told patients that we would know so much more about MS in the next five to ten years. We indeed knew so much more, but not as much as I had promised to Normie.

Normie then took us to meet his girlfriend, Linda. Linda was resting in bed watching a baseball game. As much as I hate television, I know well how important it is for people with illness, providing a crucial window on the world. Linda called him "sweetie," and Normie called her "honey." I could see how happy she was to see Normie and how she in turn comforted him by taking his hand into hers and giving a squeeze.

Outside in the bright Colorado sun we took a few pictures, sat on a bench, and talked about old times, catching up on people and what had happened to them over the years. He asked me to send him a copy of the old Super-8 movie films I'd taken when we rode together on his motor scooter in high school. Then Normie repeated his question: "Well, Big Skinny, why has it taken so long to cure this crazy disease?"

How does one explain to a patient why we haven't cured multiple sclerosis? I've tried countless times. About ten years ago I began to explain it by using the analogy of trying to fix a car that won't run, an image that everyone can easily understand. Here is a car parked on the street that won't run. Why? Here is a person with MS whose nervous system is being damaged, presumably by attacking white blood cells. Why?

I shared with him what's become one of my standard ways of explaining why MS is such a difficult mystery to solve. "Trying to understand and treat multiple sclerosis is like trying to fix a car without being able to look under the hood. We don't have enough tools to check all the parts involved in the process. If the problem with the car were that the tires were flat, then we could fix it easily. If the problem with MS were just that there were a single virus that we could vaccinate against, we would have done so long ago. But MS isn't that easy. The problem with MS is under the hood, and checking under the hood is a long and laborious process. When we finally knew enough science to open the hood, we then were confronted with all the parts of the engine. Each of the parts had to be understood, and then we discovered that there were parts within parts, and of course we have no original blueprint for how the car was built. That's why it has taken so long to find the cure," I said to Normie. "After thirty years, we still don't know enough of the parts." I thought of all the research on T cells, the white blood cells the scientific community now believes are one of the primary initiators of MS. And I thought of how in the last three decades we've discovered the way in which this amazing network of T cells both fights off infection and regulates the immune system, discoveries I both witnessed and took part in.

We gave each other a few hugs and a kiss on the cheek, and then I left. I thought of what Normie faced in the coming years. I hoped the medication he was taking would help slow the progression of his illness, and that new medications to help him would be found. Nonetheless, his life would be restricted, and there was the real possibility that he would worsen further. But with his spirit, I knew he would enjoy life as best he could, watching his grandchildren growing up, taking trips into the city, continuing to tell his jokes. Nonetheless, I could not help thinking that I could have easily been the one sitting in Normie's chair.

I had told Normie that MS is a complex disease. Its complexity expresses itself not only in the biological processes that cause damage to the nervous system but in the way different people are affected by the disease. Back in Boston, I thought of this as I examined an MS patient for whom the monster had taken a different form. Peter Stadig was a forty-seven-year-old man referred to me by a colleague from Montreal named Jack Antel, whom I had trained with at the very beginning of our careers and who now heads the MS program at the Montreal Neurologic Institute. Peter Stadig came to our MS center accompanied by his wife, and he was hobbling on two canes.

"How did your illness begin?" I asked. He looked at his wife and didn't answer right away, as if he wasn't sure. I repeated the question. "What was the first sign that something was wrong?"

"I guess it happened when I ran the New York City Marathon," he finally said. "I noticed a strange aching in my legs."

"Then what happened?"

"In the next year my legs began to give me more trouble."

"That year he missed running the marathon for the first time in five years," his wife added.

"The year after that, I noticed a limp in my right leg, and I was no longer able to play singles tennis. Then I knew something was wrong. I switched to doubles, but when the limp spread to the other leg I couldn't even play doubles."

"Did you ever have an attack?" I asked.

"No," he said, shrugging.

"I keep reading about MS patients having an attack," his wife said, "but we don't know what an attack is."

I have always been struck by the very basic questions patients so often ask the doctor about MS despite all they have read. I am happy when they do, as it gives me a chance to clarify important points about the disease and erase confusion that only makes it harder for them to contend with their illness.

His records showed that MS was diagnosed two years after he had trouble with the marathon, based on abnormalities found on the MRI scan and on a spinal tap.

I pushed to uncover the history of an attack. "Did you ever have blindness in an eye?" I asked.

"No," he said.

"Double vision or numbness across the chest?" I asked.

"No," his wife answered.

"Dizziness?" I asked.

"No."

"Clumsiness in your hand?"

"Dr. Weiner," his wife repeated, now somewhat exasperated, "we read so much about attacks in multiple sclerosis, but we don't really know what an attack is." It was clear to me why his wife was confused. So much is written about MS attacks that when patients don't have them, it makes the disease even more difficult to understand.

I explained to them that an attack is a new symptom that comes on over a few days to a week then usually subsides in four to six weeks, and that not all patients have attacks. I then examined him and found that his eye movements were normal and his arms were normal. His legs, however, were weak and spastic. When I had him walk down the hallway, he required two canes to walk twenty-five feet, and it took twenty-four seconds.

He sat back down in his chair and rested his canes against the wall. "Dr. Weiner," he said with a sigh, "walking down the hall with these canes is harder than running a marathon."

Peter Stadig has multiple sclerosis, but his type is markedly different from Normie's. He has the primary progressive form, which usually manifests as difficulty with walking and progresses slowly over the years. MS starts with the relapsing-remitting process in 85 percent of patients. After a number of years, in over half of patients, it then becomes secondarily progressive, like Normie's.

Are the two types actually different diseases? No one is really sure. The different forms of MS are not only confusing for patients, they have confounded researchers as well. The primary progressive form is the form of the disease for which we still don't have a treatment that has been shown to help.

Despite his worsening over five years, Peter Stadig had never received any treatment for his MS, something common for this form of the disease. I imagined myself in his situation, and I thought of one of my favorite quotes from Shakespeare: "Nothing will come of nothing." I knew that if I didn't try some form of treatment, there was no chance to help him. Then I thought of the Hippocratic oath: "First, do no harm." In the end, however, it was difficult for me to face a former marathoner now hobbling on two canes and not try something, and I knew a decision to give no treatment would be difficult for the patient as well. Thus I ordered a course of intravenous steroids, a common treatment for MS and one that might provide a clue as to whether he would respond to anti-inflammatory therapy. I then called Jack Antel and we discussed the use of stronger drugs, drugs that suppressed the immune system. We decided to wait because of the toxic effects of such drugs and the uncertainty of whether they would help. Perhaps he could be considered for an experimental treatment. We both lamented that the primary progressive form of MS was the most difficult to treat.

As I thought of Peter Stadig and of Normie, I also knew that not all patients with MS became disabled. The monster can be very tame. I have cared for a woman named Angela, who is fifty-three years old and has had what we call benign MS for thirty-one years. Her exam is normal. From time to time she has mild symptoms, but she has never been on treatment. Angela's first attack was at age twenty-two, when she had optic neuritis, just like Normie. Her next attack was at age thirty-three, when she had numbness from the waist down, and difficulty walking; it lasted a month, then cleared. At age fifty-three she had a third attack that caused blurred vision and fatigue.

Why in the world had her disease remained stable after all of these years? Why didn't she turn out like Normie? I realized that had we initiated treatment when she was twenty-two, after her attack of optic neuritis, and given it for thirty years, we would have concluded that the medication had kept her disease under control.

Angela had scheduled a special visit to see me because she had heard from her cousin about a new drug just approved by the FDA for MS and wondered whether she should take it. It was difficult for me to face someone who had done so well for thirty years on no treatment and tell her to begin taking injections for her MS. I obtained an MRI to make sure there was no active inflammation in her brain and then followed her closely without treatment. Angela's case made me realize even more how important it was to solve the mystery of MS and that one of the major mysteries, one of the tentacles of the monster, was that MS is a disease that can behave so differently in different people.

WHEN MY SON was eight years old, he asked me, "What do scientists do, Dad?"

I thought of possible answers: "They study how chemicals interact." "They do experiments to understand how the world works." But before I could answer, he gleefully blurted out, "I know what they do, Dad. They show slides!"

I smiled. He was right.

Although I had taken him to the lab and shown him mice and test tubes and machines that measured radioactivity, what had impressed him most vividly were the countless hours he'd watched me spend at my desk at home or at the kitchen table working on slides. When he was eight I used 35 mm slides placed in a circular carousel; now I use PowerPoint presentations on my computer.

Recently, after returning from a scientific meeting on MS, where I sat through one lecture after another, I was struck by the fact that after all these years, everyone was showing the same slide to describe MS. "Everyone is showing the same slide," I said to myself, almost out loud, as I walked along the corridor of Boston's Brigham and Women's Hospital, the same place I walked thirty years earlier when I saw John Saccone, the MS patient that piqued my interest and began the odyssey. There was less controversy about the disease, more consensus. Did that mean we finally know what causes MS? Could we be showing the wrong slide?

In 1996, when my son was twenty-three years old and had finished college, I was invited to Toronto by one of my former fellows, Marika Hohol, to give the McKewan lecture on multiple sclerosis. A car was supposed to pick me up at the airport, but by some mistake I was picked up in a big white stretch limousine that dropped me off at Marika's house in the Toronto suburbs for dinner. As I rode in the giant limo, I shuffled through the slides for my lecture. I had a set talk that I usually gave for MS, most scientists do, like the stump speech of a politician. But for this lecture, I did something different. I had become interested in physics and found myself listening to physics lectures on tape when I drove to and from work or while traveling. On one level, physics seemed so much more tangible and real compared to MS: Newton, gravity, electrons, black holes, Einstein, quantum mechanics, the big bang. Physics was full of hypotheses—hypotheses being tested and challenged, hypotheses trying to find a solution to seemingly contradictory observations, such as the integration of quantum mechanics and the general theory of relativity. Physicists were searching for the T.O.E. (the theory of everything).

What about a unifying theory for MS? How could Normie's, Peter Stadig's, and Angela's different experiences with MS be part of the same overall puzzle? When I accepted the invitation to speak in Toronto, I assumed I would take out my MS stump speech, I would load the same cassette. Then I began asking the question: Did we know enough to construct a unifying hypothesis about MS? After all, that's what the slide was about that everyone was showing, a consensus picture of the disease. But MS was too complex for a single hypothesis. There were so many questions. What started the disease? Why did it affect people differently? And, of course, a question that physics never has to confront, how to treat a disease in human beings? Nonetheless, like physics, MS needed a unifying hypothesis.

Thus, in the weeks before the talk, I began to list what I believed were the most important points about MS—all the pieces of the puzzle. Of course, the first point was, "What was it in the body that caused the damage in MS?" After I laid out the pieces of the puzzle, the next question was how to put them together, how did they fit? And what about pieces that I knew had to be there or were postulated to exist but I didn't know their shape. Could I construct a theory of MS without knowing all the pieces? I thought back on my physics lectures and I realized that if I didn't know the shape of a puzzle piece or if there was a part of the equation that was unknown, I could just make that part of the hypothesis. After all, Einstein did it when he constructed an equation regarding the universe and put in a cosmological constant, a factor called lambda. He put in lambda because he thought there was an unknown factor needed to explain why the universe was static. As it turned out, Hubble showed that Einstein was wrong about the universe being static—he didn't need lambda. I smiled to myself. If Einstein could create an equation about the universe and be wrong, I could put together a hypothesis about MS that was useful, even if parts of it were wrong too.

I ended up with twenty-one pieces to the puzzle. The twenty-one points synthesized the work from many investigators, all those on the mountain, and it contained elements from my own laboratory. In one instance I declared that something was a fact without knowing for sure. In another instance I stated that in one particular area an unknown defect existed; it was my lambda. I combined animal studies, human studies, immunological theory, and points related to treatment. Marika loved the talk and I was asked to write it up for publication in the *Canadian Journal of Neuroscience.* Since then, the twenty-one points have become my MS stump speech. I have added a video to it and I elaborate on the individual points as new information is discovered. I have given the twenty-one points to my scientific colleagues for critical review, and most agree with virtually all the points. In telling the story of the monster, I will illuminate the twenty-one points and how they have evolved over the past fifty years up until the very latest scientific thinking on the causes and potential cures for the disease. I will share with you the story of my own history of research in MS to illustrate, in a personal way, the full fits and starts involved in the pursuit.

I have placed a full discussion of these twenty-one points in a separate chapter toward the end of the book, to be explained after we have explored the story of MS. The unraveling of the truths about MS, like all scientific truths, does not happen all at once. The truths are revealed like the peeling of an onion, layer by layer as each new insight is made. This is what we will do as we tell the story of MS. Nonetheless, for those who would like a preview of where we are heading, here is a summary that includes most of the twenty-one points:

> Multiple sclerosis is caused by a white blood cell called a T cell, which goes astray and attacks one of many protein structures in the myelin sheath, which is what covers the nerve fibers in the

brain and spinal cord. We all have T cells in our bloodstream that can cause MS. There are different types of T cells: those that can cause MS, and those that serve as regulators that keep the disease-causing cells in check. Something appears to be wrong with the immune system in MS patients that upsets this balance. We don't know exactly what triggers the T cells to attack, but it probably relates to common viruses and how our immune system reacts to them. However, there is no evidence that MS is caused by a unique "MS virus," and MS is probably not a single disease but a collection of subtypes. There are different stages of MS: an early stage in which people have attacks from which they recover, and a later stage when the illness is more slowly progressive. In this later stage there appears to be a degenerative process that goes on in the brain which was triggered by the T cells. Many treatments have been shown to help MS, and the MRI scan has been a major tool in showing that treatments work and in learning about MS. In order to stop the progression of MS, we will need to give some form of treatment over long periods of time, although not every patient will benefit from every treatment. Finally, the earlier treatment for MS is started, the more likely it is to be effective.

The present state of MS research and treatment was brought into focus this past year when I was asked by a physician to evaluate on an emergency basis a thirty-year-old woman with new neurological symptoms. She was a striking woman with bright red hair who danced in musicals.

It wasn't clear to me at first why the physician had insisted that I see her immediately. The woman had had only one symptom, numbness on the side of her foot, which had lasted for only three weeks. I sat on a stool in the examining room and performed a neurological exam, testing reflexes, muscle strength, and coordination. Her exam was completely normal and only took four minutes to perform. I watched her walk. She glided down the hallway in a way that I imagined her gliding across the stage.

Nonetheless, despite her normal appearance and normal exam, I soon discovered that something was indeed wrong. Few people would have paid attention to the numbness, but the physician who referred her to me was her father, and when she told him about the numbness he had immediately ordered an MRI scan of his daughter's brain. Unlike her exam, the scan was not normal. It showed spots on her brain typical of multiple sclerosis. Her father was not surprised. A month before, his wife had died of MS, and his son, who also suffered from MS, could not walk without crutches. It is rare for people to die of MS, and MS doesn't usually run in families this strongly. Hers was a unique situation, and it made her case more urgent. She was now normal—could anything be done to keep her that way?

I stepped out of the examining room and called her father. We discussed her case in detail, and he told me about his wife and son. Finally, there was a pause in our conversation as he waited for my response.

"We have a very good chance of helping your daughter," I said finally, in a confident tone.

"How good?" he asked.

"No guarantees," I said, "but we've learned an enormous amount about MS, and we now have treatments for the disease. If we place your daughter on treatment and follow her closely, we have an excellent chance of preventing her from becoming disabled."

The father was relieved.

I must admit, I myself was relieved that I could offer hope. As I watched his daughter walk out of our MS center with the smooth glide of a dancer, I thought of Normie and his call to me for help, his confidence in me. Even though I couldn't make Normie walk again, perhaps we hadn't let him down after all. We had the twenty-one-point hypothesis and a good chance to help the dancer.

Then I thought back to 1971, when my oldest son was only one year old and my youngest son was not yet born. It was a time when there were no approved treatments for multiple sclerosis. It was a time when there was no MRI scan by which to diagnose and follow treatment effects in MS patients. It was a time when there was no Internet, no fax machines, and not even a Xerox machine. It was a time when I was a neophyte neurology resident walking the halls of the Peter Bent Brigham Hospital in Boston.

2

THE SPINAL FLUID MYSTERY

In 1971 the Peter Bent Brigham Hospital was a small teaching hospital of Harvard Medical School. The Brigham, as it is called, had been built in 1911, and in sixty years it hadn't changed much. A series of two- and three-story buildings were connected by a long corridor called "the pike." The pike was more than a passageway. Walking down the pike from one building to the next, I'd invariably meet another doctor who shared a patient with me, and we'd end up in a short consultation. In 1971 the Brigham was a homey place with serpentine stairways, open wards, and a small doctor's eating room that had a fireplace on one wall. Today the Brigham is a giant complex with a twelve-story inpatient facility, a modern outpatient atrium and foyer, an Au Bon Pain near the entrance, and a banner outside advertising it as one of the best hospitals in the country. Such marketing didn't exist in 1971 and is emblematic of the vast changes in medicine that have occurred over the years. The pike, however, still exists, and one can still get into an impromptu consultation just by strolling down it.

In 1971, just off the pike on a ward called A-Main, I stood at a patient's bedside. He had just been admitted to the hospital, and I was presenting his case to our attending physician, Dave Dawson. I was a first-year neurology resident, just beginning my specialty training, and had just admitted a patient to the hospital the day before.

"This is the first Peter Bent Brigham Hospital admission for John Saccone," I began, "who is a thirty-one-year-old man admitted to the hospital for trouble walking. Mr. Saccone owns a restaurant in the North End. He is married and has two children. There is no history of neurological disease in his family." Saccone was a short, stocky man with gray eyes, a flashing white smile, and dark hair that was already beginning to thin in the front. From his physique, one could imagine that he liked the food he cooked as much as his patrons did.

One of the medical students standing next to me began scribbling notes about the case in his tiny black notebook as I continued to read from the patient's chart.

"He first noticed trouble walking about a week ago," I said. "He thought he had sprained his ankle or pulled a muscle in his back. But over the weekend, he began bumping into the walls of his house, and he fell down. That's when his wife brought him in."

"Can he walk now?" Dave Dawson asked.

"Only if two people hold on to him," I said. "It's like he is really drunk."

"Does he have any weakness in his legs?" Dawson asked.

"His legs are remarkably strong," I said.

"What did you find on neurological exam?"

"I didn't find that much, just ataxia of all his limbs."

At this point John Saccone perked up. "What's ataxia?" he asked.

"It's loss of coordination," Dawson said, placing his hand on Saccone's shoulder.

Dave Dawson was in his mid-thirties, a handsome man with an angular face and an easy smile. He had received his neurology training at the Boston City Hospital, was a junior staff member at the Brigham, and was an outstanding clinical neurologist. He had spent time in the lab studying muscle disease but gave it up for clinical neurology. And although he had played guard on the football team during his undergraduate years at Harvard, he was not aggressive or threatening; he was direct, someone who never minced words. Residents and students liked him and trusted him.

"Which area of the brain is damaged if you have ataxia of all the limbs?" Dawson asked, turning to the medical student.

"Cerebellum," the student said.

"Do you agree, Dr. Weiner?" he asked me.

"Yes," I said.

"And what do you think is going on in his cerebellum?" he then asked.

I hesitated for a second. My first thought was of a tumor of some kind, but I didn't want to say that in front of Saccone. So instead I just said, "I'm not sure."

Dawson then performed a neurological exam on Saccone himself, looking into his eyes and checking the speed and rhythm of his eye movements. Then he tested the strength and coordination of Saccone's arms and legs and sensation. Finally he tested Saccone's reflexes and stroked the bottom of one of his feet with his key. By his manner, I could tell Dawson had learned something from this simple exam that I had missed, but I couldn't tell what it was.

"Mr. Saccone," Dawson asked immediately, "did you ever lose vision in one of your eyes?"

"Now that you mention it, I did go blind in one of my eyes when I was in high school," Saccone responded, evidently a bit surprised that Dawson somehow knew to ask the question. "No one ever knew what it was, and my eye came back to normal. The doctor told me it was a viral infection of some kind."

"Did you ever have tingling in your hands or feet," Dawson asked, "even slightly?"

"No," Saccone said.

"Did you ever bend your head down and feel something like electric shocks down your back?"

"No."

"How about funny feelings in your belly? Did you ever have sensations of a band or a constriction around your waist?"

"Yes, I've had that," Saccone said.

"How long ago?"

"Three years. It lasted about a month and I never paid much attention to it, but my pants and my pants buckle felt funny."

With a nod, Dawson motioned for me and the medical student to step into the corridor.

"He doesn't have a cerebellar tumor," Dawson said to me. "What does he have?"

"What did you find on exam?" I asked.

"The exam doesn't matter," Dawson said.

"I know," I muttered under my breath, feeling a bit foolish for even asking the question. I had immediately realized what the patient had when Dawson elicited his history of blindness, and I was irritated with myself for not asking about that. One of the basic tenets in getting a medical history from a patient is to start with a hypothesis about the diagnosis and to test it by asking specific questions. Too often the patient leaves crucial things out. I had been in a rush to examine John Saccone because there was a case waiting for me at the time in the emergency room, and I hadn't formulated a hypothesis, but I knew exactly what his blindness indicated.

"Multiple sclerosis," I said.

"How do you know that?" the medical student asked.

"Multiple events in time and space," I said, repeating what I myself had learned as a student. "In other words," I continued, "MS affects multiple places in the nervous system. Most neurological diseases are localized to one place in the nervous system. But MS is in many places—it affects the eyes, it affects the brain, it affects the spinal cord. That's exactly what Dawson uncovered in John Saccone's history. Think about it," I said to the student. "Three places in the nervous system were involved. Blindness—obviously the eye. Trouble with coordination—the cerebellar part of the brain. And the bandlike sensations across the abdomen are a sign of disease affecting the spinal cord."

The student wrote down everything I said in the little book he carried in his coat pocket.

"Second, there were multiple events in time that occurred over a period of years. He had blindness in high school, bandlike sensations across his abdomen three years ago, and now the trouble walking."

"But they were so spread out in time," the student said.

"Classic for multiple sclerosis," I said, drawing on what I had read, for this was the first time I had seen a diagnosis of MS made at the bedside.

"What causes MS?" the student asked.

At that Dawson laughed. "Figure out what causes MS and how to treat it and you'll win a Nobel Prize."

The student managed a smile but wrote nothing in his notebook.

"Did you see anything when you looked into his eyes?" I asked Dawson. "Could you see blurring of the optic disk or paleness of the disk that can be seen with optic neuritis?"

"Let's take a look," Dawson said as he led us back to Saccone's bedside. Saccone looked apprehensive, and I began thinking of how we would break the news to him. Dawson handed me an ophthalmoscope so I could examine John Saccone's eyes. "See if you can tell which eye it was."

I asked John Saccone to stare straight ahead as I shone the light on his pupil. John Saccone's gray eyes were even more striking because of his dark hair and a one-day growth of dark beard he had accumulated since coming to the hospital.

The pupil is actually a clear hole through which one can look into the eye. It appears black because there is no light inside the eye. By shining light through the pupil, one illuminates the back of the eye and can see the head of the optic nerve; it is like standing outside a darkened house and shining a flashlight through the living room window. That's why with flash pictures the eyes appear red—the light from the flash is illuminating the retina on the inside of the eye, which is red.

I rapidly clicked on the lens of my ophthalmoscope to focus on the optic nerve. The optic nerve is one of the most dramatic sights in medicine and is not easy to see the first time one tries. I remembered vividly the time I first looked into an eye to try and see the optic nerve. "Do you see it?" my instructor had asked. "I think so," I'd said hesitantly. My instructor had laughed. "If you only think so, then you haven't seen it."

As I steadied John Saccone's head with my hand so I could peer into his eyes, I felt the sweat on his forehead and realized how much anxiety he must be feeling. His right optic nerve came into focus, looking like a bright golden sun in a red sky. Blood vessels spread out like rays from the center of the sun, and if I looked carefully, I could see them pulsing. The optic nerve is a direct extension of the brain and a common site of attack in multiple sclerosis. I could discern no abnormalities of the right optic nerve, but when I looked at the left one, I immediately saw that something was wrong. The golden yellow sun was pale, especially on its perimeter. "It's the left optic nerve," I said. Dawson nodded, and I confirmed it by measuring visual acuity and finding that Saccone's visual acuity was 20/60 in the left eye and 20/20 in the right eye.

"What's going to happen to me?" John Saccone asked nervously. He explained that he had two boys, ages seven and three, and his wife was expecting a third child. She worked as a teacher but would take a prolonged leave of absence after the birth of their third child.

"You'll get better," Dawson said immediately, not inviting further questions because he didn't want at this time to tell Saccone that he probably had MS.

"What type of treatment will you give me?" Saccone asked.

"We'll discuss that after we get the results of all your blood tests," Dawson said, "and I want Dr. Weiner to perform a spinal tap to make sure you don't have an infection. Then we'll talk about treatments." Dawson didn't actually think Saccone had an infection, but it was easier to explain the spinal tap in that way. The spinal fluid might actually show signs of multiple sclerosis and was a key test to diagnose the disease.

John Saccone nodded hesitantly, realizing he had no choice but to wait.

After we left Saccone's room, the student asked about what treatments there were for MS. Dawson explained that the only treatments available were oral steroids or intravenous adrenocorticotropic hormone (ACTH), a hormone that stimulated the adrenal gland to produce cortisone. The treatments were given over two to three weeks, but people with the relapsing form of MS usually recovered from an attack without any treatment.

"Will he recover faster if he is given the steroids?" the student asked.

"One of the few controlled studies in MS, done six years ago, showed that patients given three weeks of ACTH recovered faster from their attacks than those given a placebo, but there was no long-term follow-up to see what happened afterward."

That's a problem, I thought. I understood Dawson's reasoning that Saccone most likely would recover from his current attack. But what if he had further attacks and became more disabled? This was my first direct encounter with MS, and I was struck by how limited the treatment options were and that the only real laboratory test we had for MS was the spinal tap.

It had been discovered in MS that there was an elevated level of antibodies called gamma globulin in the spinal fluid. Thus, finding raised levels of gamma globulin in the fluid obtained in a spinal tap often helped in the diagnosis. The cause of the increased gamma globulin in the spinal fluid of MS patients is unknown, but over the years many have thought that solving this mystery holds the key to understanding the disease.

Spinal taps are also done to diagnose infections of the brain, since there are increased numbers of white blood cells in the spinal fluid during infections. If the infection is caused by bacteria, as happens in meningitis, white blood cells called neutrophils accumulate. If the infection is caused by a virus, as occurs in encephalitis, white blood cells called lymphocytes accumulate.

Lymphocytes are the major cells of the immune system, fighting off infection. But they also are involved in other situations. For example, lymphocytes are responsible for rejecting a transplanted kidney if the tissue match isn't appropriate. It is also known that during an MS attack, there may be a slight increase in the number of lymphocytes in the spinal fluid, even if there is no infection to be found. No one knows for sure what draws the lymphocytes to the spinal fluid.

I turned Saccone onto his side and had him draw his knees to his chest and grab his knees with his hands. This caused the bony prominences in his lower back to stick out and made it easier for me to guide the needle into the spinal canal. I remembered how nervous I had been when I did my first spinal tap as a medical student. I palpated the prominences with the tips of my fingers, covered his back with a sterile sheet, and then cleaned the area with a disinfectant. Finally I took out a four-inch needle, angled it below one of the bony prominences, and with one swift movement thrust it into his back. I pulled out the stylet in the middle of the needle, and crystal clear fluid began dripping out, filling three separate tubes. I had become good at doing spinal taps.

After I was done I told John Saccone to lie flat in bed for the next few hours so he would have less of a chance of getting a headache from the spinal tap.

As I helped him turn over in bed, he grabbed my hand. "Can I talk to you, Doctor?" he said.

"I have to take your spinal fluid to the laboratory," I said, reluctant to get into a discussion until we had the results of the spinal tap.

"I need to talk to you now," he insisted.

I knew a few minutes made no difference in the analysis, so I pulled up a chair and sat at John Saccone's bedside; that way we could talk comfortably while he lay on his side.

"I'm scared," he said. "What's *really* going to happen to me?"

I didn't know what to tell him. I wanted to comfort him, but I didn't want to give him wrong information. Furthermore, he hadn't been told that he had MS, which was Dawson's responsibility. I assumed Dawson was right and John Saccone would recover from his attack without treatment. "The body can naturally heal itself," I said, "and the chances are excellent that you'll recover from this attack." I told him Dawson needed the results of the spinal tap to make a final diagnosis and that I wanted to avoid speculating on a diagnosis. Then we discussed whether he would need to take medication, and he asked whether anything else could be done to help him recover.

I thought for a second. "We can give you physical therapy and gait training," I said. When John Saccone didn't seem convinced, I added, "I'm sure that will help you recover," even though I wondered how it was possible for me to be sure.

Saccone grabbed my hand. "I hope you're right, Doctor."

I squeezed his hand, glad that I could comfort him even in this small way. Then I grabbed the tubes containing his spinal fluid and made my way to a tiny laboratory in the basement of the hospital. With a small pipette, I placed a single drop of his spinal fluid on a counting chamber and covered it with a glass slide. Immediately I could tell there was an increase in the number of lymphocytes in his spinal fluid; I counted sixteen cells per microscopic field, whereas normal was less than four cells. There in that small basement lab, I sat for a moment and tried to imagine what was happening in John Saccone's nervous system. Why were there increased numbers of lymphocytes? Were they searching for an unknown virus lurking in the brain? Were they attacking the brain, as they did in rejecting a mismatched kidney? Or were they a secondary event, reacting to something else that caused his attack? The second tube I sent to the laboratory for gamma globulin determination, and the third tube I sent to the bacteriology laboratory to check for infection. The results came back just as Dawson had predicted. There was no evidence of infection, but there was a marked elevation in the level of gamma globulin, which confirmed the diagnosis of MS.

The next morning on rounds Dave Dawson, myself, and the medical student stood at John Saccone's bedside once again. Dawson explained the results of the spinal tap and finally told Saccone that he had multiple sclerosis. He explained that we didn't know what caused MS but that we believed it was somehow related to a virus and that Saccone would almost certainly recover from his attack. The only treatment for attacks was the ACTH, which was given intravenously or by injection.

John Saccone seemed conflicted. He didn't want foreign substances in his body, but he was afraid he might not improve on his own. So Dawson told Saccone that he'd start him on physical therapy and watch his progress carefully. If he didn't show signs of improvement, then the next stage would be treatment with ACTH.

John Saccone agreed, and his care then became my responsibility. I would give him a thorough evaluation every morning before rounds. Later, at lunch in the doctors' dining room, Dawson told me that if John Saccone didn't show signs of improvement within three days, he would order the ACTH. He didn't want Saccone to feel that nothing had been done for him, although it was clear in Dawson's mind that because MS was a chronic disease that progressed over fifteen to twenty years, one course of ACTH would not have a major impact on the long-term outcome of Saccone's disease.

The first day that I performed a neurological exam on John Saccone, I made one of my first clinical discoveries about treating MS, though I didn't really see it as such at the time. I had created a special page in his chart to monitor his progress, with columns for each part of the neurological exam, which I filled in precisely. When I finished the exam I realized that the nuances of all the results were not as important as one simple measure of his disability: how he could walk. I had John Saccone walk down the center of the ward and timed him. Patients in beds on either side of the ward served as spectators for the event. It took him thirty-five seconds, and he needed the use of a walker.

"When do you test me again?" Saccone asked.

"Tomorrow morning before rounds," I said.

"Is it cheating if I practice?" he asked.

"You can practice as much as you want," I told him.

When I tested Saccone the following morning before rounds there was no change in the amount of time it took him to walk down the center of the ward, and he still required the walker. But as I watched him push the walker in front of him, I thought I could discern improvement; his legs did not seem as wobbly as they had the day before. And on the second day, I was almost certain that his coordination was better. Although Saccone didn't realize it, the third day was crucial, because if there was no sign of improvement by day three, Dawson was ready to start the ACTH.

On the third day Saccone's improvement was obvious. He pushed his walker along with smooth steps, and it took him only eighteen seconds to cross the ward, almost half the time from when I started measuring him. Dawson was not surprised.

Twelve years later we would publish an article in the *New England Journal of Medicine* in which we introduced the ambulation index, a simple measure of disability in MS based on how the patient walked. It would become widely used by physicians evaluating MS patients. Without knowing it, I was performing a rudimentary ambulation index on John Saccone.

"When can I go home?" Saccone asked.

"When you can walk with a cane," Dawson said.

John Saccone was overjoyed. As I watched Dawson test Saccone's reflexes, I imagined looking through Saccone and inside his brain and spinal cord. MS is a disease that primarily affects the insulation surrounding the nerve fibers in the central nervous system. That insulation is called the myelin sheath. The sheath is wrapped around the nerve fibers like the insulating material that covers a telephone wire, and in MS it is swollen, surrounded by lymphocytes, and being eaten away by scavenger white blood cells called macrophages. Damage also occurs to the underlying nerve fibers, which are called axons. I wondered: was Saccone improving because the insulation in his brain and spinal cord was repairing itself? Or was it a simple matter of the swelling going down? Maybe that's what Dawson was thinking that made him so sure John would recover on his own. Unless he had another attack, the swelling would go down and he had to improve.

What about the lymphocytes and macrophages surrounding the myelin sheath, cells that presumably were causing the damage? I had seen such cells with my own eyes under the microscope. Were they dying? Were they being killed off by the body's protective cells, acting to shut down the myelin-damaging cells? I would have loved to be sitting on top of the cells I saw under the microscope and get a firsthand look at what was happening in John Saccone's nervous stem that allowed him to walk better.

The next day Saccone could walk the distance in less than fifteen seconds, and on the following day he did it with a cane in ten seconds.

Dawson discharged John Saccone on the weekend, in time to attend his son's baseball game. Saccone left the hospital using a cane, but when he returned to the neurology clinic for follow-up three months later, his walking was virtually normal.

After John Saccone was discharged, I began to think more and more about multiple sclerosis. What process could come and go in the central nervous system, leaving a person blind in an eye or unable to walk for a while, and then clear up so the patient was virtually normal? Clearly, the body was using its own mechanisms to shut off the disease process. If we could only understand how the body did it, we could shut off the disease process ourselves. The biggest clue had to be the inflammation in the brain. I had seen the increased number of white blood cells in John Saccone's spinal fluid as well as the raised level of gamma globulin in the spinal fluid, and I'd become intrigued by the mystery of why they were there. The thought struck me that perhaps understanding the elevated gamma globulin in the spinal fluid was the key to the disease, because it was an immune reaction against something; we just didn't know against what.

I immediately went to the library to read about all that was known at that time about the raised gamma globulin in the spinal fluid. I knew that gamma globulin is produced by the body to fight off infection. In the library, I read that gamma globulin is an antibody shaped like the letter Y. The tip of the Y attaches to a virus or bacteria and neutralizes it. Such antibodies are very specific. For example, polio vaccination induces antibodies that react with the polio virus but not with the measles virus. Indeed, I found that day that most of the known causes of elevated levels of gamma globulin in spinal fluid are infections. Syphilis is one of the classic diseases that causes raised gamma globulin in the spinal fluid, and there were early reports that MS was caused by the same infectious agent that causes syphilis.

Elevated levels of gamma globulin in MS were first described in 1942 in a paper by Elvin Kabat. No one knew what causes gamma globulin to increase in MS, though it was obvious to postulate that there was some sort of infection in the brain of MS patients. An infection would explain both the raised gamma globulin and the increased white blood cells I had seen so clearly in John Saccone's spinal fluid. The body's immune system uses both cells and antibodies to fight off infections. T cells (so named because they come from the thymus gland) patrol the body and directly attack infectious agents. Antibodies are much smaller than T cells, are dissolved in the serum, and are made by immune cells called B cells (named after the bursa of Fabricius in the chicken; there is no such bursa in humans, but they are still called B cells in humans). This was all well and good, but if there was an infection in the brain of MS patients, what was it? To date, no one had discovered an infectious agent in MS brain or spinal fluid, though there had been numerous attempts.

I then found papers about a crucial series of experiments demonstrating that the increased gamma globulin in MS spinal fluid was unique and not simply derived from the gamma globulin we all have in our bloodstream. There are enormous amounts of gamma globulin in everyone's bloodstream, and the big question these experiments were trying to answer was whether the gamma globulin in the spinal fluid of an MS patient has leaked into the brain and then into the spinal fluid from the bloodstream or is actually produced within the brain itself. To answer that question, a group of doctors in New York had injected gamma globulin that was tagged with a radioactive marker into the bloodstreams of MS patients. They then performed spinal taps to determine whether that gamma globulin made its way into the patients' spinal fluid. If the tagged gamma globulin they injected appeared in the spinal fluid, they would surmise that the gamma globulin was leaking in from the bloodstream. If not, they would have to conclude that it was being produced in the brain. Few experiments give clear-cut results, and too often the results are subject to many interpretations. In this instance there was no question: the increased gamma globulin was being synthesized within the brains of the MS patients.

The next question was obvious. What is it in the brain that causes the abnormal levels of gamma globulin? In other words, what is it in the brain that the gamma globulin is reacting against? Researchers had concluded that the cause must be either an infection, probably a virus, or an abnormal immune reaction against the brain tissue itself. The cells I saw under the microscope could be associated with both possibilities.

I made a simple diagram to outline the two possibilities: (1) gamma globulin reacts with an infectious agent in the brain, or (2) gamma globulin reacts with the brain tissue itself. I then left the library and walked outside, past the entrance to Harvard Medical School and toward one of the back entrances to the Peter Bent Brigham Hospital. I could smell the salty ocean air, so different from the dry mountain air of Colorado, where I grew up and where I went to medical school. I had done some research as a medical student, and now I began to think about doing research on multiple sclerosis. This was such a mysterious disease, but also one for which we had a major clue. I wondered whether with a little more research, that mystery about whether an infection or some kind of problem with the immune system was the culprit could be solved. It also seemed to me that this was a disease that might be relatively easy to treat. After all, patients could get better on their own.

3

IS MS TWO LETTERS OR THREE?

When I joined the research effort to find a cure for MS, it was known that there are increased numbers of white blood cells in the brains of MS patients, but no one knew why. There were two main theories: a viral infection in the brain or an abnormal immune response. Research was being done to look closely at the brain to identify the processes at work, and that research produced vivid pictures of the damaged brain tissue in MS patients. The damage was clustered around the cavities known as ventricles, deep in the brain where spinal fluid is produced. From there the spinal fluid circulates, bathing the entire brain and spinal cord. It was not known why the damage clusters around the ventricles. Also, at this time, research was focusing on what was believed to be the animal version of MS. Research on animals was being used to test whether MS was caused by an abnormal immune response, and if so, how that abnormal response could be controlled. My next step in learning about MS was to attend a conference devoted to MS and the latest research findings.

The conference was being held across town at Massachusetts General. The expert who would be discussing the topic was Raymond Adams, the Bullard Professor of Neuropathology at Harvard Medical School. In 1961 he wrote a classic paper describing in detail how the brain of an MS patient appeared under the microscope. I would get to look under the microscope with him at tissue from a patient with MS. What better way to learn about the disease?

Massachusetts General Hospital was far more imposing than the Brigham. Whereas the Brigham was a series of three-story buildings connected by the pike and had hospital beds for 250 patients, Mass General consisted of three towers, one fifteen stories high, and had beds for 1,000 patients. It was also more famous. Some of its fame came from the special clinical pathological conferences (CPCs) that were published every week in the *New England Journal of Medicine.* At a CPC, a case of a patient who had been treated in the hospital but died was presented to an expert. The expert made a diagnosis based on the clinical history and then found out if the autopsy confirmed the diagnosis. The proceedings of the conference were then printed in the *New England Journal of Medicine* for every doctor in the world to see. As students we had competitions to see who could make the right diagnosis.

Massachusetts General Hospital was called MGH, Mass General, or simply the General. Some of the residents called it the WGH, the World's Greatest Hospital. In 1971 the Brigham and the General were crosstown rivals. There was little communication between the two, and the training programs were different. Any doctor, of course, was welcome to attend a conference at MGH, but I must admit I felt a bit strange taking the trolley across town. I took the Green line trolley from in front of the Brigham, changed downtown at Park Street, and then took the Red line to Charles Street, next to the Charles Street jail. In 1997, buses shuttled across town between the two hospitals and in 2005 the Charles Street jail was being converted into a luxury hotel attached to the Mass General.

Back in 1971 the neuropathology conference was held on the first floor of the Warren Building. The sign on the door read Neuropathology Conference, but we neurology residents called it "brain cutting." The brains were neatly sliced and placed on a white pan. Nearby was a microscope where one could examine the cells inside the brain. When I entered the room, I didn't recognize any of the doctors, but I did notice a difference between the staff doctors at MGH and our hospital. They wore short white coats; staff doctors at the Brigham wore long white coats.

I would have liked to sit in the front row, but I knew that row was reserved for the professors and senior staff doctors, so I sat as close as I could, in the second row. Although I had never seen Dr. Adams, I assumed he was the doctor sitting in the very center of the first row.

The conference began with one of the neuropathology residents presenting the case. "This is a thirty-two-year-old woman with multiple sclerosis," he began. "She is a unique case, since she died in a car accident and we were able to examine her brain during the early stages of the disease. Her course was typical, with optic neuritis followed by numbness and tingling in her legs and then difficulty with gait. At the time of her accident she was ambulating with a cane." The resident looked up from his notes and asked if there were any questions. There were none.

"The general pathological examination was unremarkable," he continued. "This is consistent with multiple sclerosis, which affects only the brain and spinal cord but no other parts of the body. The brain on gross examination was of normal size and weight."

The resident then projected a 35 mm slide of the brain that had been cut in half and showed the clear delineation between the gray matter on the surface of the brain and the white matter underneath. Deep in the white matter were the ventricles, which appeared as cavities in the brain. Simple description, I thought to myself as I looked at the slide: the gray matter looks gray and the white matter looks white.

"Classic areas consistent with multiple sclerosis were observed surrounding the ventricles," he said as he pointed out areas where the white matter of the brain was discolored and abnormal. He then outlined other areas in the white matter not next to the ventricles where there was also damage to the brain caused by multiple sclerosis. The slide was a graphic illustration of the lesions or scars made in the brain by MS, and over the years I would introduce my own talks on multiple sclerosis with a copy of this slide. Indeed, thirty years later I would show the slide when I gave the twenty-one points talk at a symposium on MS in honor of Dr. Adams.

"I have a question for you," one of the doctors asked. "Why do you think the lesions of multiple sclerosis surround the ventricular cavities? It looks like there is something leaking out from the cavities into the substance of the brain and causing the damage."

I immediately thought of the elevated levels of gamma globulin in the spinal fluid and wondered whether it leaked from the cavities into the white matter, causing the damage.

At this point one of the doctors in the front row spoke. He spoke in a measured, authoritative voice. It was Dr. Adams. "I don't think the scars around the ventricular cavities are caused by something leaking from the spinal fluid into the brain," he said. "Rather, there are veins surrounding the cavities of the brain, and white blood cells exit from these blood vessels and damage the surrounding white matter."

The resident thanked Dr. Adams for his comments and then projected a series of slides that depicted the appearance of multiple sclerosis under the microscope. The slides showed that there were many white blood cells in the brain, cells that had infiltrated the brain from the bloodstream. If one looks at a normal brain under the microscope, one can see the nerve fibers, their myelin sheaths, and brain cells, but there are no white blood cells. In the brain of someone with MS, there are many white blood cells, and these cluster around the blood vessels in the brain, like ants clustering around spilled sugar. Because of the white blood cells, MS is classified as an inflammatory disease. The central question remained why these white blood cells were infiltrating the brain. Were they there to fight off a virus, or were the white blood cells themselves causing damage to the brain tissue or reacting to brain damage caused by something else?

After a discussion of multiple sclerosis and a recent scientific article on it, people were invited to look under the microscope at MS tissue and examine the actual brain slices of the MS patient discussed at the neuropathology conference.

I jumped at the chance and joined Dr. Adams in looking through the double-headed microscope at the slides. He pointed out to me the white blood cells that had infiltrated the brain and the damage to the myelin sheath, which was accompanied by macrophages, or scavenger cells, that had engulfed the myelin debris. I then touched the brain slice with my fingers and could feel that the areas involved by the disease were firmer compared to normal tissue. They were sclerotic—they formed hard multiple scars, hence the name multiple sclerosis. I had directly confronted multiple sclerosis with the tip of my finger.

When I told Dr. Adams of my interest in MS, he introduced me to a doctor in a short white lab coat named Barry Arnason, who was performing research on the autoimmune theory of MS. Arnason invited me to his laboratory on the third floor of the Warren Building, two floors up from the neuropathology conference, to discuss MS research.

Arnason was a man in his late thirties, short with piercing blue eyes and a high-pitched, often irreverent voice. He smiled easily and was happy to learn that I was interested in MS. His laboratory was a small labyrinth of four interconnecting rooms, and we sat in a tiny office just off one of the rooms. Arnason's office was wonderfully cluttered with boxes of microscopic slides piled on top of each other, books on immunology and multiple sclerosis, and manuscripts at various stages of completion. I noticed a picture of him at the ocean with his wife and son and another picture of him sitting at the microscope with two other doctors, one of whom looked like a younger Dr. Adams.

We briefly discussed the MS case presented at the conference and reviewed the two major theories of the cause of MS, a viral infection or an autoimmune response. While he talked, he intermittently wrote with his left hand on a tiny blackboard that was on his door to summarize the scientific points he was making.

"I think MS is caused by the immune system attacking the brain—an autoimmune response," he said. "In fact, we are trying to set up a system in the lab to study lymphocytes and discover what structure the lymphocytes are attacking in the nervous system." I knew that lymphocytes were white blood cells that made up one of the major limbs of the immune system.

Arnason went on to explain that one of the hallmarks of the immune system was how precise it was when it attacked viruses and protected us against infections. We all knew that after exposure to chicken pox, the immune system could attack and fight off the chicken pox virus but not the mumps virus. In MS, there was only damage to the white matter in the brain and spinal cord, no damage to any other part of the body. If MS was caused by the lymphocytes of the immune system attacking the brain, what was it in the brain that they were attacking?

"If we can figure out what structure the lymphocytes are attacking," Arnason said, "we might be able to shut them down and cure the disease." This of course assumed that the lymphocytes were the cause of the disease.

Arnason went on to explain that one of the components of the myelin sheath in the brain that was a prime candidate was a protein called myelin basic protein, or MBP, and his laboratory was carrying out research on MBP. He then asked whether I was interested in coming to the lab for an elective period during my residency and learning about MS research.

The thought of doing research on MS excited me. Here was a chance to do something that could change people's lives so much more than giving steroids to John Saccone.

Arnason had watched me take notes on everything he said. He sat back in his chair and twirled the piece of chalk between his hands. "Howard," he said, "if you are serious about getting into MS research and are interested in the immune theory of the disease, then you must learn about EAE, experimental allergic encephalomyelitis—the best animal model of MS we have based on the immune theory of MS. There is very little you can do in three months in the lab, but you can do an experiment in EAE."

He then outlined the experiment. He wanted me to prepare MBP from a human brain and use it to treat guinea pigs with EAE. If the experiment worked, I could use the MBP I had prepared to test lymphocytes from MS patients to see if they reacted to it.

I took it both as a challenge and a reward. If I could do a successful animal experiment, I could then test the same principle in humans.

Next Arnason gave me an assignment. "Before doing your experiment, I want you to present the classic paper on EAE written by Thomas Rivers to our laboratory journal club." Arnason grabbed a piece of paper from his desk and began scribbling. "When you present the Rivers article, divide your presentation into these four parts: background, hypothesis, experimental results, and conclusion." He handed me the paper. "That's how science is done," he said.

I took the piece of paper and slid it into my notebook. I was back in school again. I immediately went to Mass General's library and luckily found the article by Thomas Rivers in the dusty stacks in the basement. I read it on the trolley back to the other side of town and underlined everything I didn't know with a red pen. I was overjoyed, but there were a lot of red marks.

THE STORY OF MS is indeed the story of EAE as well, because many researchers have argued that EAE is essentially the same disease as MS but in animal form. That argument has spurred enormous debate among scientists over the decades, who continue to argue the degree to which the disease process that occurs in EAE is related to the disease process that occurs in MS. The study of EAE is based on the theory that MS is an autoimmune disease in which the immune system attacks the brain. Advances in our understanding of the immune system have immediately been applied to the treatment of EAE, and if they are successful, they are considered as potential new therapies to be tested in MS patients. Furthermore, EAE has been studied in hopes of uncovering key defects in the immune system that might be linked to the cause of MS.

Animal models such as EAE are crucial for developing treatments for human disease because animal experiments can be done faster and under controlled conditions. Larry Steinman, a major figure in both the study of EAE and the immunology of MS, is a close colleague and friend, as we both went to Dartmouth College, we both married women from Israel, and we both devoted our scientific career to the study of MS. He is now professor of neurology at Stanford. Once as we drank coffee during a break at a scientific conference in Germany, he said to me with a smile, "Howard, if I don't come up with a new treatment for EAE every six months, then I'm not making progress in the lab." Studies in the EAE model have led to treatments that have been shown to be effective in MS, although there are treatments that worked in the EAE model but failed in MS. To this day the majority of treatments that are tested in MS are first tested in the EAE model.

The story of EAE began in 1933 with the experiments of Thomas

Rivers, and that's why Arnason wanted me to study his classic experiments and present them at the laboratory meeting. The journal club was held during lunch on Tuesdays in a small conference room just off the lab. At the journal club there were postdoctoral fellows who worked in the lab, research technicians, and neurology residents like myself who were doing an elective.

I followed the outline Arnason had given me and divided my presentation into four parts, each part introduced by a new overhead. The nervousness I felt when I began disappeared as soon as I became caught up in the presentation.

Background

The first description of EAE is credited to Rivers, with his classic experiment published in 1933 and a follow-up paper in 1935. Rivers was a famous virologist, and his experiments were carried out to explain a disease that occurred after rabies vaccination and in people convalescing from certain infectious diseases, especially smallpox and measles. Surprisingly, the brains of these individuals didn't show signs of a typical viral infection; rather, they showed a loss of myelin and the infiltration of large numbers of white blood cells, very similar to what happens in MS.

Hypothesis

Rivers wanted to know what it was in the rabies vaccine that was causing a reaction in the brain that resembled MS. He hypothesized that there were two possibilities: (1) the rabies virus in the vaccine or (2) the brain tissue used to grow the vaccine.

The question Rivers was asking was a critical one. If he could produce an illness in monkeys similar to MS just by injecting them with brain tissue, it could mean that MS was an allergic reaction to brain tissue, and one would then have a model for MS that could be used to understand the disease and devise treatments.

Experimental Results

In Rivers' first experiment, he injected monkeys with one of three compounds: a lab stock of rabies virus, brain extracts taken from normal rabbits, or an extract from nonbrain tissue. The monkeys received three inoculations a week for over ten weeks directly into the muscle.

The results were clear-cut. Changes similar to MS were seen only in monkeys injected with the brain extract. However, only two of the eight monkeys injected with brain extract developed these changes. Thus he carried out a more extensive second set of experiments two years later: Eight monkeys received brain extract injections, and eight monkeys were left untreated. This time each monkey received up to eighty-five injections, and the injections were given over an entire year. The results were even more dramatic than the first experiment. Six of the eight monkeys that received the injections developed clinical disease, whereas none of the control monkeys became sick. Furthermore, when the brains were examined under the microscope, there was inflammation and demyelination in the brains of seven of the eight monkeys who were injected with the brain extract, but none in the control monkeys.

When I read the papers, I thought of John Saccone. The animals had trouble with their eye movements and had ataxia, or abnormality of their gait, just like John. In addition, Rivers found both white blood cells and an increase in gamma globulin in the spinal fluid of the sick monkeys, just as I had found in Saccone.

Conclusion

Rivers concluded after his second series of experiments that brain tissue injected into animals could cause an allergic reaction that looked similar to MS. Thus, the animal model of MS was born: experimental allergic encephalomyelitis. The word *encephalomyelitis* means there was inflammation affecting both the brain and spinal cord. Recently, some scientists have replaced the word "allergic" with "autoimmune," but the letters EAE remain.

Rivers' experiments have become classic. The next major breakthrough occurred in 1947, when Elvin Kabat was able to use a special adjuvant, or compound that boosted the immune response, to induce EAE in guinea pigs after one or two injections rather than eighty-five injections.

Arnason complimented me on my presentation and briefed me on the experiment I would perform with MBP. If it worked in EAE, I would then test MBP in MS patients.

The letters "MBP" are almost as famous in MS research as the letters "EAE." MBP stood for "myelin basic protein," one of the proteins that was part of the structure of the myelin sheath. It was called "basic" as opposed to "acidic" because of its chemical charge. In the years since Rivers conducted his experiments, scientists had discovered that the MBP in Rivers' injections could trigger the EAE. Many therefore thought that MBP held the key to understanding and treating MS because scientists postulated that in patients with MS the immune system attacked the MBP in the myelin sheath, causing damage. We now know that other proteins in the myelin sheath besides MBP can serve as a target of the immune system (point two of the twenty-one points). Furthermore, it was felt that the attack on the myelin sheath was carried out by the T cells of the immune system.

In 1962 Arnason had performed an experiment showing that one couldn't get EAE in animals if there were no T cells. As I've noted, T cells get their name from the fact that they mature in and obtain their biological properties from the thymus gland, just below the neck. They then leave the thymus gland and populate the rest of the body, where they function as a crucial part of the immune system not only in fighting off viruses but as cells that regulate the immune system. With no thymus gland there are no T cells. Arnason's experiment was to remove the thymus gland in animals just after they were born, so they had no T cells. Animals without T cells were completely resistant to EAE.

The experiment Arnason wanted me to perform was one that he had been thinking about for over a year. He needed someone to make human MBP so he could test the hypothesis that MS patients had increased numbers of T cells that were targeted to MBP. My experiment was to prepare human MBP and to demonstrate that it was biologically active by injecting it into guinea pigs and showing that they became sick with EAE. Then I would take my MBP preparation and test whether there were increased numbers of T cells in MS patients that were targeted to MBP by mixing the MBP I had made with blood samples from MS patients.

The preparation of MBP was, as we say in the laboratory, "cookbook": no new concepts, no theories to be tested, just follow the instructions. I obtained a sample of normal human brain from the autopsy room and dissected out the white matter, which contains the myelin that is attacked by white blood cells in MS. Within the myelin are lipids and a large number of proteins, one of which is MBP. I took the white matter through a series of steps to isolate the MBP. First I extracted the lipids with alcohol, and then I used different sugar concentrations to separate proteins from the brain. The final step was to place the solution that contained MBP over a large cylindrical column of filtering beads. I placed the column on the laboratory bench in front of a window overlooking the Charles River. It was almost four feet high, and I had to stand on a chair to put the solution on top of the column. A large number of proteins ran through the column, and the MBP was separated from the other proteins according to size and was collected in a single test tube. I then ran a gel to show the protein was pure.

Armed with the gel, I sat again in Arnason's little office to learn how to design the EAE experiment in guineas pigs. "It's very simple," Arnason said. "You are going to inject the MBP to make sure it is biologically active, that it can cause EAE. If it can't cause EAE in animals, then we can't test it in people. In order for it to cause EAE, you must put the MBP in an adjuvant and make an emulsion." An emulsion is a mixture in which tiny droplets of liquid are dispersed in another liquid but remain separate droplets instead of dissolving.

"What if I inject it and it is not in an emulsion?"

"It won't work," Arnason said. "In fact, it could protect the animals."

"You mean the exact same protein that causes the disease could protect against the disease?" I asked incredulously.

"That's right," Arnason said. "If it is not in an adjuvant and made into an emulsion, it could turn off the T cells that cause EAE rather than activating them."

"Can I do that as part of my experiment?" I asked. "Can I try to treat the animals by injecting MBP that's not in an adjuvant and emulsion?"

Arnason took a piece of chalk from his blackboard, twirled it in his hands, and thought for a moment. "Why not?" he said finally. "You'll be testing the biological activity of the MBP in two ways: first, how well it induces disease, and second, how well it protects against disease."

He stared out the window at the Charles River and collected his thoughts. Then he turned to the blackboard and wrote the design of the experiment. The experiment would be performed using ten guinea pigs divided into two equal groups. After injecting the animals with MBP in the adjuvant emulsion to induce EAE, the first group would be injected with MBP that was in a solution to determine if such treatment prevented EAE. The second group would be injected with albumin, which served as an irrelevant or control protein to demonstrate that injecting a protein that wasn't MBP didn't protect the animals from EAE.

Martha, one of the technicians in the laboratory with experience in EAE, helped me with the experiment. Although I knew enough from my chemistry background to prepare MBP, I would need help with the EAE experiment.

I took the clear solution containing the MBP I made and divided it into two test tubes. One tube I would use to induce EAE, and the second I would use to prevent it. To induce EAE, I needed to mix the MBP with an adjuvant. We now know that in order to trigger T cells, other cells called macrophages are needed, and the adjuvant turned on the macrophages. The adjuvant was called "complete Freund's adjuvant," named after a Hungarian-born American immunologist named Jules Freund. Complete Freund's adjuvant contains oil, water, and killed bacteria, usually mycobacteria, which stimulate the macrophages. (Incomplete Freund's adjuvant doesn't have the bacteria.) Rivers did not use an adjuvant, and this is why it took so many injections for him to obtain EAE in his monkeys. The discovery by Kabat that with complete Freund's adjuvant one could induce EAE by a single injection and get animals sick in fourteen days was a major advance in the study of EAE and MS. Charlie Janeway Jr., a prominent Yale immunologist whom I've had the privilege of collaborating with on EAE experiments in genetically altered mice, has called Freund's adjuvant the immunologist's "dirty little secret." Without an adjuvant, it's virtually impossible to induce and study immune responses in animals.

I put the oily yellow adjuvant in one syringe, the clear MBP solution in another syringe, and connected the syringes. I then mixed the solutions by pushing them back and forth between the two syringes until I had a thick white emulsion. Martha told me that if the emulsion wasn't white and thick, it wouldn't work. It took almost thirty minutes of mixing until it was white and thick enough. We then tested the MBP to prove that it was in an emulsion by putting a drop of the mixture in a beaker of water. If it was truly an emulsion, it would float. We succeeded. We then grabbed gloves, more syringes, and the test tube with the clear solution of MBP and headed into the basement to the animal room.

In the animal room we found ten guinea pigs, one to a cage, with Arnason's name on the cages. Next to Arnason's name I wrote "Weiner/EAE." We numbered the cages and then injected each of the animals under the skin with the white MBP emulsion while they were under anesthesia. I held each animal and Martha did the injecting. We then took animals one through five and injected a clear solution containing the control protein albumin. In animals six through ten we injected a clear solution containing MBP. The animals tolerated the injection well and we headed upstairs.

Later I sat in a tiny cubicle assigned to me and wrote out the details of the experiment. Then, as Arnason had asked me, I wrote down the potential outcomes and headed to his office. "Write them on the board," he said to me, settling back in his chair.

I took a piece of chalk and copied from my notebook.

"Perfect results," I wrote, "animals 1–5 get sick, animals 6–10 are protected."

"Right," Arnason said. "That means the MBP you made is active both in inducing EAE and in protecting against EAE."

Then I wrote, "Busted experiment: no animals get sick."

"Right again," he said. "That means something is wrong either with the MBP you made, with the emulsion, or with the animals."

Then I wrote, "All animals get sick."

"That means the MBP you made can induce EAE but can't protect," Arnason said.

Finally I wrote, "Animals 1–5 are protected, animals 6–10 get sick."

Arnason laughed. "That probably means you mixed up the tubes you used to inject the animals."

Martha and I checked the animals every day, and I came in on the weekends. When twelve days had gone by and no animals showed signs of getting sick, I began to worry about a busted experiment. Then on day thirteen, the animals in cages two and three developed difficulty walking. By day fifteen all the animals in cages one through five had trouble walking. The animals in cages six through ten looked normal. Moving from concern to elation, I began to imagine a perfect experiment. On day seventeen, all the animals in cages one through five were paralyzed, and those in cages six through ten were normal.

Amazing, I thought as I adjusted the food and water for the paralyzed animals. I had encountered firsthand the ability of MBP to cause an immune reaction in animals that caused paralysis. What would Rivers have thought? He had had to inject monkeys for a whole year to get EAE. Even more amazing was that as I continued to follow the sick animals, they began to recover. By day thirty, the animals in cages one through five were perfectly normal. My experiment, however, was only semiperfect. On day twenty-one, the animal in cage seven became paralyzed, and animals six and ten developed mild trouble walking. Animals eight and nine, however, were completely protected. Thus the MBP solution had lessened the incidence and severity of EAE but didn't completely suppress it.

At the end of the experiment I tested the animals for the presence of T cells that reacted to MBP. If there were T cells that reacted to MBP and I mixed them in a test tube with MBP, they would be stimulated and would grow. I could measure the growth by adding a radioactive label called thymidine, because as the cells divided they would take up thymidine. Increased counts of thymidine meant the cells were growing. The results were clear. In the animals with EAE the radioactive counts were between 35,000 and 40,000. In the animals without EAE the counts were between 1,000 and 2,000. Thus, I had directly demonstrated that injecting animals with MBP in adjuvant had expanded the numbers of MBP-reactive T cells.

I thought again of John Saccone. What had happened to the animals mimicked very closely an acute attack of multiple sclerosis, in which patients recover just as John Saccone did. The animals were paralyzed and then walked normally. John Saccone also ultimately recovered from his attack and walked normally. But EAE was an experimental animal disease, and MS was MS. Now that the EAE experiment had worked, I was ready to use the MBP I had made to test whether there were increased numbers of T cells in MS patients that were targeted to MBP.

The EAE experiment I performed wasn't exploring uncharted territory. We knew what the results should have been, and if the experiment didn't work, I could repeat it over and over, making small adjustments each time, until it finally did work. The MS experiment I was about to perform was different. We were testing a hypothesis that predicted certain results, but if I didn't get those results, it was possible that our hypothesis was wrong. If our hypothesis was wrong, the experiment would never work, no matter how many times I did it or how many variables I changed in an effort to make it work. The hypothesis I was testing was central to understanding MS. If MS was like EAE and EAE was caused by an allergic reaction to MBP, I should be able to find increased numbers of T cells that reacted to MBP in MS patients.

I would test for T cell reactivity to MBP in MS patients using the same assay I had used in the EAE animals. I would take the human MBP I prepared, mix it with lymphocytes isolated from the blood of MS patients, and measure incorporation of radioactive thymidine. If there were T cells in the blood that were sensitized to MBP, they would be stimulated and would grow. I knew the human MBP I made was good because it caused EAE and activated T cells from animals with EAE.

Instead of going downstairs to the animal room, I now went to the hospital to draw blood from MS patients, for which I had to ask permission. The first person I drew blood from was a woman in her thirties who was in the hospital recovering from an attack without any treatment, just like John Saccone.

"Why are you drawing my blood?" she asked.

I explained that we were trying to find the cells that caused MS and told her briefly about the EAE experiment and how we were now testing the theory in MS patients. She was excited by the project and, like so many MS patients, was enthusiastic about helping out with such experiments.

"How long before we have a cure?" she asked.

I was just beginning in MS research and didn't know what to say, so I ignored her question as I put the tourniquet around her arm and prepared to draw her blood.

"I read somewhere that we should have a cure in ten years," she said.

"I hope so," I said weakly.

"So you think there is a good chance?" she asked.

"Yes," I answered almost reflexively as I plunged the needle into her vein.

Over the years it has not become that much easier to answer the question. The path from basic discovery in the laboratory to clinical trials to a cure is a long and tortuous one, but every experiment is a step toward the cure, even if the experiment fails. Thus there is the delicate balance of being optimistic but not giving the patient false hope.

As the blood billowed into the syringe, I imagined her bloodstream as a river penetrating into every part of her body. Blood moves through the body at enormous speeds. Inject something into the bloodstream and in ten seconds it's everywhere. The bloodstream is comprised of red and white blood cells; red cells carry oxygen and white cells are used by the body to fight off infection. The blood contains trillions of white blood cells, and I was trying to identify the specific T cells among them that, during the course of their journey, left this woman's bloodstream, entered her brain, and attacked the myelin sheath around her nerves, causing her attack. In the guinea pigs, I had created such myelin-attacking T cells by injecting MBP in adjuvant. Did such T cells exist in MS patients, and were there more of them in patients with MS compared to people who didn't have MS?

I drew blood from five MS patients that day and five patients who had other diseases affecting the nervous system. The patients with other diseases affecting the nervous system served as controls; I needed to prove that increased reactivity to MBP was observed only when the nervous system was affected by MS, not by other diseases. Then I isolated the lymphocytes and divided them among twenty tubes, using two tubes for each patient. I added MBP to one of the two tubes. If there were T cells that reacted to MBP and the cells grew, there would be increased radioactivity in the tube to which I added MBP compared to the tube without it. I let the tubes incubate for seven days, then collected the cells and counted the radioactivity in a special scintillation counter. A perfect experiment would be increased counts in the five MS patients and no increased counts in the non-MS patients. As when I checked for reactivity to MBP in the guinea pigs, the results of the experiment came as a printout from a scintillation counter that measured radioactivity.

I found myself standing in the closet at the end of the hallway where the scintillation counter was located. Many investigators had the printouts brought to their desks, where they analyzed them. I couldn't wait, so I stood next to the counter as each of the tubes was analyzed for radioactivity and the results printed tube by tube. The counter was set to spend thirty seconds counting every tube, and my heart began beating faster as the numbers began to appear. Maybe I would have beginner's luck and find increased counts in the MS patients. That would be a breakthrough. If the MS patients had increased reactivity to MBP, maybe they could be treated with MBP, just like the guinea pigs. I struggled to read the numbers as they were printing. When the counting was over, I tore off the sheet and examined the numbers closely. There were twenty numbers, two numbers for each patient. The results couldn't be clearer.

There was no difference whatsoever between the MS patients and the controls—all the numbers were between 1,000 and 1,500. I felt a pit develop in my stomach. Everything had gone so right with the EAE experiment that I had unconsciously assumed my experiment on MS patients would work.

I put the tubes back in and this time counted each tube for five minutes, hoping to amplify a positive effect. Still no difference. The counts had increased only because of the additional amount of time spent counting; they were all now between 10,000 and 15,000.

I made my way to Arnason's office and showed him the results. Although a bit disappointed, he didn't seem that surprised by the numbers. When we discussed what to do next, he emphasized that what I had done was just a pilot experiment. There was so much more work to do in the laboratory and more variables to test before we would know whether or not there was reactivity to MBP in MS patients.

In retrospect, I realized there was no way I could discover anything of significance about such a complex disease as MS by working in the lab for three months. What I did do, however, was take the plunge. With my simple EAE experiment in guinea pigs and my abortive attempt to find reactivity to MBP, I joined my fellow scientists and embarked on my odyssey.

My attempt to find T cell reactivity to MBP in MS was an experiment that investigators would continue to perform in a variety of ways for the next quarter century. Fifteen years later, in my own laboratory across town at the Brigham and Women's Hospital, I would be performing a far more sophisticated variation of it myself.

Before finishing my rotation, I went back and reread Rivers' experiments. I discovered that as part of Rivers' experiment to induce EAE in monkeys, he also tried to find a virus in the animals. Thus Rivers was addressing the very first point of the twenty-one-point hypothesis: is MS caused by a virus or by a T cell that initiates an allergic reaction against brain tissue? In the 1933 Rivers experiment, he tried unsuccessfully to transmit the disease from one monkey to another as a test for a virus or other infectious agent. Thus, his experiments not only created the EAE model but foreshadowed attempts to transfer MS from people to animals in a search for a virus or other infectious agent. Rivers was cautious about his inability to find a virus and wrote, "Failure on our part to find an infectious agent does not necessarily mean that one was not present."

Absence of proof is not proof of absence.

The same could be said of my failed attempt to find T cell reactivity to MBP. Just because I didn't find it didn't mean it wasn't there. Of course, it didn't mean that it *was* there either.

My three-month rotation was over, and I would return to the hospital to discover that not all MS symptoms were as simple as T cells attacking the myelin sheath. As I spent more and more time with MS patients I would learn just how complex the effects of the disease were, in both scientific and human terms. The increased gamma globulin in the spinal fluid would be investigated by many researchers over the next three decades, and although it would come to be accepted that it represented an increased immune response in the brain, not a reaction against a virus, the exact structures it reacted against would remain a mystery.

In one of my discussions with Arnason, I wrote a simple equation on his small blackboard: MS = EAE. Was MS two letters or three? That is, to what extent was MS identical to EAE? As it turned out, although most researchers felt MS was more than just EAE, the very first point of the twenty-one-point hypothesis was that, just like EAE, MS was initiated by a T cell that reacted with a structure on the myelin sheath, and it initiated damage by going from the bloodstream into the brain after it was activated. Viruses were believed to be the main culprits that activated the T cells, but they were not responsible for the damage.

"Well, what do you think?" Arnason asked me on the day I left the lab. "Is MS two letters or three?"

"I have no way of knowing," I said. "Many more experiments need to be done."

"You've learned a lot in three months," he said.

4

THE ELECTRICITY IS OFF

One of those MS patients who helped so much to open my eyes to the true complexities of the disease, and whom I will never forget, is a woman named Norma Mason, whom I treated for a short time after I finished my three months doing research on MBP. I had become more and more fascinated with MS; I read as many research articles as I could about MS and kept a special watch over any MS patients who were admitted to the neurology ward when I was the chief resident. To help follow the large number of patients I was responsible for, I kept three-by-five cards on all the patients but would put a star next to the name of any patient who had MS.

Norma Mason was admitted to the hospital late one afternoon because of increased difficulty walking and problems with urination. The bladder is connected to the nervous system at the level of the lower spinal cord, and patients with MS often have bladder problems. She was forty-two years old and had had MS for almost fifteen years. She needed a cane to walk and was admitted to the hospital because she had recently started having more trouble walking. In today's era of managed care, we could no longer admit such a person to the hospital for observation, but in the past it was a luxury we could afford.

I was now chief resident, in the last year of my training, and was supervising a first-year neurology resident. The resident took a history, examined the woman, and wrote a detailed note in the chart. I sat at the patient's bedside and read through the note to familiarize myself with her case. The resident had joined me for my evaluation of the patient.

Norma Mason was divorced and had no children. She was fiercely independent and worked in sales traveling around New England even after her MS began to cause problems. Her first attack was at age twenty-five, when she had optic neuritis and went blind in her left eye. She recovered from her attack and was normal until age thirty, when she had numbness that began in her feet and went up to her waist. She had trouble walking, and a diagnosis of MS was made after she had a positive spinal tap. Over the next ten years she had more attacks, and two years before, she had begun using a cane because of weakness on the right side and some spasticity.

I looked at the resident and then at Norma Mason, who watched intently as I read her chart. I had one of my three-by-five notecards with her name on it and jotted down pertinent parts of her history. Later, I would add her notecard to my packet of notecards with all the patients I was responsible for on the ward and, because she had MS, I put a star in the corner of her notecard as I had begun to save all the MS cases I saw.

I turned to the patient. "How dependent are you on the cane, Mrs. Mason?" I asked.

"I can walk without it, but not very far," she said.

It wasn't clear to me whether she was having a major attack like John Saccone's, or whether the difficulty walking reflected the slow worsening of her disease. Her condition had been changing slowly over the past few months, not abruptly, like John Saccone's. I decided to watch her closely over the next twenty-four to forty-eight hours and order laboratory tests. The major question in my mind was whether we would administer a course of ACTH while she was in the hospital.

I sat with the resident at dinner and we talked about MS. I told the resident of my time in Arnason's lab and my experiment with the EAE mice. We then talked about the mechanism of MS attacks and how an attack could be caused by the movement of lymphocytes from the bloodstream into the brain, where they attacked the myelin sheath. I liked to teach.

When I was a first-year neurology resident, Larry Levitt, who was the chief resident at the time, and I published a handbook called *Neurology for the House Officer.* Now I noticed that the resident had our handbook in the pocket of his white hospital coat. In 1974 there were only two pages on MS in our handbook. They described the common clinical symptoms of MS as part of a chapter that dealt with increased reflexes in young people. Many MS patients had increased reflexes because of involvement of the spinal cord. In 1974 there was no mention of mechanism of the disease or treatment. Thirty years later, in the seventh edition, there would be a separate chapter devoted to MS, the theory of what caused it, and a list of treatment options.

After I finished our evening rounds, I made my way up a serpentine staircase to the third floor of the old Peter Bent Brigham Hospital. I read some articles, reviewed my notecards, and fell asleep just before midnight.

At 3:30 a.m. I was awakened by a frantic call from the resident.

"Something's happened to Mrs. Mason," the resident said. "She is totally paralyzed. She can't move her arms or legs!"

"What do you mean she can't move her arms or legs?" I asked in disbelief, still half asleep.

"Just what I said," the resident repeated, "she's totally paralyzed."

"Is she breathing?" I asked. "What's her blood pressure?"

"She's breathing fine, and her blood pressure is normal," the resident said.

I was shocked to hear this about Norma Mason. How could she have deteriorated so rapidly? But as I made my way from the on-call room, I began to think about what might be happening to her, and by the time I arrived at her room I had a hypothesis.

Norma Mason indeed could not move her arms or legs, and although her breathing was normal, I found she was having difficulty swallowing.

"I already called the pharmacy," the resident said, "to order ACTH and steroids, which we can give to her intravenously." But as I surveyed the situation, I wasn't so sure she'd need them.

I began my exam by asking Norma Mason to follow my finger with her eyes, which she did with no trouble. There was panic in her eyes, though, something that could not be quantified as part of the formal neurological exam, but it was clearly there. I lifted up her legs. They were like dead weights as they fell back onto the bed. She could barely squeeze my hand. This was a woman who had walked down the hospital ward with a cane less than twelve hours earlier. I was profoundly impressed by the strangeness of this disease.

Not only were Norma Mason's frightened eyes staring at me, I could feel the stare of the nurse and resident, waiting for my next move. I put my hand on her forehead and then on her cheek as I began to test my hypothesis.

"What's her temperature?" I asked the nurse.

She looked down at the chart. "It was ninety-nine when we took it four hours ago," she said.

"I think she has a fever," I said. "Let's take her temperature again."

The nurse put a thermometer under Norma Mason's tongue and held it there for three minutes. When she withdrew it, it read 103.7°.

"That's why she can't move her arms and legs," I said. "She has a fever." If she had a fever, then I knew she must have some kind of infection. I listened to her lungs to see if that was where it was. But her lungs were clear, and there had been no vomiting or diarrhea to suggest an infection in the gastrointestinal tract.

"Let's get a urine sample," I said. "I bet she has a urinary tract infection."

We used a catheter to get a clean urine sample, and we checked the urine in a small laboratory off the hospital ward. Sure enough, it turned out that Norma Mason had an acute urinary infection. Even more, we later learned from blood cultures that the infection had spread to her bloodstream. That's why she had the rise in body temperature.

"She doesn't need steroids," I said to the resident. "She needs antibiotics and a cooling blanket." This was my first direct experience with the effects that fever can have on an MS patient. When a person with MS gets a high fever, the already damaged nervous system can shut down.

By noon Norma Mason's temperature had come down, and the paralysis in her arms and legs and her trouble swallowing disappeared just as quickly as they had appeared.

What happened to Norma Mason that night was not an MS attack. An MS attack is caused by the movement of white blood cells from the bloodstream into the brain and usually develops over days, not hours. Instead, she had experienced an acute electrical shutdown of her nervous system caused by the rise in her body temperature, which had made it even more difficult for her damaged nerve fibers to conduct electricity. Her attack was temperature-dependent and easily reversible.

Both the brain and the heart need electricity to function. In the heart there is a bundle of fibers that carries electricity from one chamber to the next and synchronizes the heartbeat. In the brain the movement of electricity is far more complex and the electricity travels at enormous speeds. If I decide to wiggle my finger, that decision triggers an electrical impulse in my brain. The electrical impulse begins in the cortex, the thinking part of the brain, and moves through the brain, down the brain stem, and into the spinal cord. The electricity leaves the spinal cord and enters the nerves in my arm, which in turn trigger the muscles that cause my finger to wiggle. In a similar fashion, when light strikes my eye it triggers electrical impulses that travel from my eyes into nerves (called the optic nerves), which then join nerve fibers that course through my brain until they reach the back of my head (the occipital cortex), and that is where I see. The optic nerves are especially vulnerable in MS, and that's why inflammation of the optic nerve (optic neuritis), which causes blindness, is so common in the disease.

The basic structure that carries electricity in the brain and spinal cord is called an axon, which can be thought of as an electrical wire. Just as electrical wires have insulation around them, so do axons, and that insulation is none other than the myelin sheath. The myelin sheath is required for the function of the axon, and without it the axon does not conduct electricity well. Multiple sclerosis is classified as a demyelinating disease. In other words, the myelin or insulation surrounding the nerve fiber is damaged, and it is that damage to the myelin sheath and the underlying axon that causes the host of clinical symptoms MS patients suffer: numbness, tingling, and shocklike sensations down the back. When I reviewed my three-by-five cards, I had accumulated a remarkable list of symptoms reported by MS patients: "When I walk, my feet feel like I am walking on sand." "My hand feels like a piece of wood." "I feel like I have a tight belt just underneath my breasts that goes around to the back."

Actually, the electricity that travels along the nervous system does not travel on the axons themselves. The myelin sheath, or insulation, that covers the axon has breaks at regular intervals. Rather than moving along the axon itself, the electricity jumps from one break in the myelin to another. The breaks in the myelin are called nodes of Ranvier, named for the French scientist who described them. The jumping of electricity from one node of Ranvier to another is called saltatory conduction, and it is nature's way of conducting electricity at enormous speeds over long distances, such as from the brain down the spinal cord and to the nerve.

One way to imagine the saltatory conduction of electricity in the nervous system is to imagine traveling from Boston to Los Angeles in a car or an airplane. By car, one moves on highways and it takes days to cross the country. If one flies from city to city, it takes hours. Electricity that travels in the nervous system by saltatory conduction is like a plane flying from city to city; the electricity jumps from one of the breaks in the nerve to the next. But if the myelin is damaged, the electricity can no longer jump, and it moves along the axon at a snail's pace. Norma Mason experienced rapid deterioration in her nervous system function because the process of saltatory conduction is extraordinarily sensitive to temperature. If the temperature goes up in a person with MS, this conduction of electricity is slowed. If the temperature goes down, the conduction is enhanced. This temperature effect doesn't happen in people without MS because the myelin sheath is not damaged. Thus Norma Mason's attack was not an attack in the classic sense of white blood cells damaging myelin; it was the temporary loss of function of the nerve fibers caused simply by a change in body temperature. After her bladder infection and infection in the bloodstream were brought under control, she passed her general medical evaluation and left the hospital in the same condition as when she had entered. She was later given treatment with steroids, but it did not help significantly.

The exacerbation of symptoms of multiple sclerosis by changes in temperature has been known since the end of the nineteenth century, when a German physician named Uhthoff described the phenomena as it related to visual symptoms. Subsequently, the heat sensitivity of the nervous system in MS patients led to the "hot bath test." In the early years, when it was unclear whether a patient had MS, doctors sometimes placed patients in a hot bath to artificially raise their body temperature. The purpose was to determine whether signs and symptoms of multiple sclerosis could be brought out by a hot bath. Many people with MS notice that their symptoms worsen when they take a hot shower or are at the beach on a hot day. Nonetheless, there is no compelling evidence that temporarily raising the body temperature is damaging to the nervous system of an MS patient, even though many patients are told by their doctors to avoid hot temperatures. Exercise raises body temperature; and MS patients may experience an increase of symptoms when they exercise and become overheated, even though exercising is generally good for MS patients.

If raising the body temperature makes MS symptoms worse because an axon without a myelin covering doesn't conduct electricity as well at higher temperatures, one could postulate that lowering body temperature would make MS symptoms better. This indeed is the case. Some MS patients find that they walk better when they come out of a cold swimming pool than when they went in. Unfortunately, when their body temperature reequilibrates after thirty minutes, they are back to their pre-swim status. Some have tried to sell MS patients cooling vests to help neurological function, but these vests have little practical applicability, even though a study by NASA confirmed that lowering body temperature can have temporary positive effects in people with MS.

One of the most sensitive ways to measure the flow of electricity in the nervous system is to measure the amount of time it takes for an electrical signal to travel from the retina to the back of the brain after a special light pattern has been flashed in front of the eyes. This test is called a visual evoked response test, and before the MRI was developed, this test was frequently used to diagnose MS by demonstrating that more than one part of the nervous system was involved. Dr. W. Ian McDonald of London has spent much of his career studying the physiology of demyelination and amongst his many contributions, found that when MS patients drink ice water, he could find improvement in vision and in the amplitude of the visual evoked response.

Electrical short-circuiting can take other forms, and one of the classic symptoms of multiple sclerosis is called Lhermitte's sign, in which MS patients experience a feeling similar to electric shocks down the spinal cord when they bend their neck. The electric shocks are due to damage to the upper cervical spinal cord; bending the neck stretches the spinal cord and causes the sensation of electric shocks. In addition, some patients have different levels of energy during the day that may relate to electrical short-circuiting.

MS patients who experience these temperature effects can sometimes figure out ways to regulate their temperature to temporarily reduce the untoward effects. I once had an MS patient who wanted to get his wife pregnant but was having trouble maintaining an erection because the MS affects the spinal cord (this was before the days of Viagra, which has benefited some MS patients). He told me that after he heard a talk on the effect of temperature on MS he devised a plan. Before making love to his wife, he took an ice-cold shower. The lovemaking was cold, but at his checkup, he showed me a picture of their new baby boy.

Changing body temperature is not a true cure for impaired nerve conduction, but drugs are being tested that may help a demyelinated axon conduct electricity better. Such drugs may help not only with heat-related symptoms but also with other symptoms. The only problem is that after white blood cells attack the myelin sheath, axons may become severed and unable to conduct electricity irrespective of the temperature.

One of the most frustrating features of MS is for patients to understand the cause of their symptoms. They hear that MS is a disease in which the white blood cells attack the myelin sheath and yet there are symptoms related to conduction of electricity that have nothing to do with white blood cells. They hear about MS "attacks" and that people recover from attacks, but some patients never have attacks and not all patients recover from their attacks. And then, there is a symptom that is not only confusing to patients but frequently can fool the doctor. It is a symptom rarely discussed in the written material given to MS patients. This is another type of MS symptom that many think is an "attack" and that also relates to electricity, but in this instance the electricity is "on," not "off." Although it can be extremely disabling for the patient, it is easily treatable. Just as Norma Mason had taught me so much about what happens to MS sufferers when their electrical conducting is reduced by temperature effects, another patient helped me to understand this other kind of attack.

Glenda Kinney came to see me because her MS was "on" even though neither she nor her doctor knew it. Glenda Kinney was a grade-school teacher with two small girls of her own. She was also an artist and had set up a special place in her home studio where her girls could dabble on their own canvases when she painted. She was sent to our MS center by her doctor because she was having frequent MS attacks.

At our MS center, we have a sheet that patients fill out on which they list how they have been doing since their last visit. Although MS is a complex disease and there are countless nuances in understanding how the disease is progressing, the patients are asked to simply check off a box that tells how they feel their MS is compared to the last visit: improved, unchanged, or worse. I am always relieved when I see that a patient has checked off "improved," though the improvement may be due only to the fact the patient has recently recovered from an attack, and I worry that it is only a matter of time before there will be another. I most like to see "improved" in someone who has started a new therapy, or "unchanged" checked off for many visits in a row though some patients check off "unchanged" even when they are actually a bit worse.

Glenda Kinney had checked off the "worse" box on her form. In the space below the box, she wrote "In the past month I have been having many MS attacks."

Glenda Kinney was in her thirties and had had the relapsing-remitting form of MS for five years. Like many MS patients, she had had an attack of optic neuritis that caused temporary blindness in her left eye. She had also had two other attacks, numbness in her waist and an attack that affected her ability to walk, for which she was treated with steroids. The steroids appeared to have a dramatic effect, and both her walking and her neurological exam returned to normal.

As I watched her walk from the waiting area to the exam room, I noticed no problem with her walking or coordination, so I knew she wasn't having a major attack affecting motor function. In fact, it seemed strange to me that she looked so well in the face of what she had written about having so many attacks. Perhaps I would find something on exam.

I sat on a stool facing her.

"My MS was fine until two months ago, when I began having more attacks," she said. "My doctor gave me treatment with steroids, but it didn't stop them."

"How many treatments with steroids have you received?" I asked.

"Two," she said. "One a month ago, and one last week."

"Did the treatments help at all?" I asked.

"No. If anything, the attacks are worse."

I was puzzled. "Tell me about your attacks," I requested.

"The attacks affect my ability to speak," she said. "I get slurring of my words, and I also get dizzy when it happens. When the steroids didn't help me, I took short-term disability from school. I just can't teach anymore. I even find it hard to paint. I never know when I will have an attack." She then told me her last attack had occurred that day, while she was sitting in the waiting room, and had lasted about a minute.

"How many of these attacks do you have a day?" I asked, smiling to myself. Based on what she'd told me, I was almost certain I knew exactly what was happening to her.

"It depends," she said. "Sometimes as many as twenty a day."

Now I was convinced I knew what was causing her attacks.

I performed a neurological exam.

Normal.

A normal exam in someone who had just had an MS attack.

Then, as chance would have it, she had an attack just as I finished the exam.

"I'm having one now," she said with fear in her voice. "There's nothing I can do to stop it."

I asked her to follow the movement of my finger with her eyes. Then I asked her questions to check her speech. Soon she was slurring her words, she felt dizzy, and she had difficulty walking. There was little to find on exam during the attack apart from some jerkiness to her eye movements, and it was all over in two minutes. She told me she felt groggy afterward.

"I never know when these MS attacks are going to hit," she said. "Isn't there anything you can do?"

"You are not having MS attacks," I told her. "True MS attacks come on over days and last four to six weeks, not minutes. You are having something called tonic spasms, caused by short bursts of electricity from scarred areas in the brainstem."

Tonic spasms are a form of seizure, but unlike general epileptic seizures, in which people lose consciousness, tonic spasms in MS patients occur in an area of the brain that doesn't affect consciousness. The damaged nerve fibers in MS patients are more irritable, and that may lead them to fire uncontrollably. Fortunately, these mini-seizures usually are completely controlled by a commonly used anticonvulsant drug, carbamazepine.

"I have medication that will stop your attacks," I said confidently.

"Really?" she said. "The steroids didn't do anything."

When I saw Glenda Kinney's excitement, I wondered if I had been too confident. The medication was good, but few medicines are 100 percent effective. I didn't want to raise false hopes in a disease whose course is so difficult to predict. Nonetheless, within a few days her attacks disappeared, and the following week she was back in school teaching and painting in her studio.

Glenda Kinney's tonic spasms and Norma Mason's electrical shutdown exemplify the complexities of MS, a complexity that makes it confusing for both the patient and the doctor. I later had an MS patient come to see me who was in an MS support group with Glenda Kinney and wanted the "magic pill" I'd given to Glenda Kinney hoping it would make *her* MS attacks disappear. Unfortunately, she did not have tonic spasms and there was little I could do to help her.

Another of the fundamental characteristics of MS that makes it complex for both patient and physician are the relapses and remissions that people experience. Furthermore, even though the brain can adapt and cope with a certain amount of damage, allowing a patient to recover from an attack, there may be permanent hidden damage that exists in MS of which both patient and physician are unaware. This hidden damage is best illustrated by one of the most common MS symptoms, blindness in one eye. As we know from Normie's case, during the course of

MS a person often goes blind in one eye due to damage to the optic nerve. This damage is caused by white blood cells that attack and infiltrate the nerve. The damage to the optic nerve can be measured by the time it takes for electricity to travel from the retina to the back of the brain, as measured by the visual evoked response test. For this test, a pattern is flashed in front of the patient's eye, and the time it takes for electrical impulses to hit the back of the brain are measured. In normal individuals it takes a few thousandths of a second.

Most MS patients who have optic neuritis recover in four to six weeks, after which their vision returns to near normal. Nonetheless, even though the patient has almost normal vision, the amount of time it takes for electricity to travel from the retina to the back of the brain as measured by the visual evoked response is still slowed. Why? We now know that other pathways in the brain take over to help restore vision. This has been shown by functional brain imaging, in which one can identify which part of the brain becomes activated when a person does a task, reads, or even thinks a certain thought. A person who has recovered from optic neuritis and has normal vision uses more parts of the brain to see than someone who has never had optic neuritis. In other words, the nervous system can compensate for damage to a certain extent, it has a certain amount of plasticity. However, with further damage, it becomes harder and harder for the nervous system to compensate, and the patient may lose some function. This helps explain another feature of MS: early on, patients may recover from attacks, but with additional attacks they accumulate deficits.

Thus, the ability of the nervous system to compensate may lull the patient and the physician into a false sense of security. Although the patient may seem to be doing well clinically, the nervous system is being taxed, and as damage begins to accumulate, the nervous system may no longer be able to compensate and function normally. Ironically, Glenda Kinney's tonic spasms, which were very debilitating, do not result in nervous system damage. Thus, electrical short-circuiting and the progressive loss of the plasticity of the nervous system are major tentacles of the monster.

We now know that after an attack on the myelin sheath by immune cells such as lymphocytes and macrophages, recovery of neurological function involves three processes. First, the inflammation caused by the immune cells is cleared and swelling of the myelin sheath goes down. Second, as we have discussed here, the brain adapts to the damage by using other brain regions to compensate. Third, the body attempts to repair the damage by making more myelin and by reorganizing some of the electrical channels in the nerve so they can conduct impulses better. Nonetheless, although some repair does occur, it is usually incomplete.

Despite the complexities in the ways MS affects different people, researchers have struggled with the fundamental question of whether any one cause could be the culprit in all cases. Is all MS caused by a single virus? Or by a disorder in the immune system? As I became more and more familiar with the vast range of experiences those with MS endure, I became determined to participate in finding key answers to help solve this basic riddle. As for all MS researchers, doing so would mean I would have to commit to a particular line of investigation; for me it was exploring either the virus or the immune system hypothesis, at least at the start, as at that time the story of the search for the cure was divided into these two main streams of exploration.

5

VIRAL ORIGINS

FOR A NUMBER of years I have cared for a woman from Texas with multiple sclerosis who is an astronomer. Her husband is also an astronomer. She gave me a copy of a book she edited about quantum mechanics, which I admit I could not understand. Nonetheless, I have always been fascinated by origins. When the astronomer comes to Boston for her periodic checkups, I take the opportunity to pose my latest questions to her and her husband about the origins of the universe. After discussing the sun, earth, and moon, the conversation invariably leads to the Big Bang. That's where I get stuck. My mind just can't comprehend an event such as the Big Bang, which came from nothing. Maybe that's where God comes in. Then our conversation always turns to the origins of her MS. *MS has to be easier than the Big Bang,* I think.

Her MS causes her fatigue, and she has occasional numbness in her feet that bothers her when she walks. Nonetheless, she works hard at her academic pursuits and tracks her MS with the same meticulous approach with which she studies the universe. She often comes to her visit with a three-page numbered list of symptoms and questions listed in categories (1.0, 2.0) and subcategories (1.1., 2.1).

"Why did I get MS?" she asks.

"What triggered it?" her husband asks.

"Is MS a disease of modern civilization?" she asks.

Origins are important. They provide clues. In the case of MS, where was the monster born and how does it come to life?

MULTIPLE SCLEROSIS was first described pathologically in the early part of the nineteenth century by Jean Cruveilhier in Paris and by Robert Carswell of London, who also worked in Paris. There is a debate over who was first with their drawings of the MS scars, or plaques, scattered throughout the brain and spinal cord. Was it Cruveilhier in 1841 or Carswell in 1838? For the astronomer, I doubt it makes much difference. A recurrent theme in scientific discovery is that two individuals independently discover the same thing at approximately the same time. There is a time for everything under heaven, even scientific discovery. At the Salpêtrière, a hospital in Paris, Jean-Martin Charcot initiated a serious study of the disease in the last three decades of the nineteenth century, distinguishing multiple sclerosis from Parkinson's disease. He correlated the pathological picture with clinical observations, and observed the disease at close hand by studying a maid in his house who had MS, with symptoms that came to be known as Charcot's triad (jerky eye movements, hand tremor, and altered speech patterns). Charcot, however, was pessimistic that treatment could be found.

One of the first clinical descriptions of MS is contained in the diaries of Sir Augustus d'Este, a grandson of George III who was born in 1794. Like my boyhood friend Normie, he had optic neuritis in his twenties and over the next twenty-six years had recurrent attacks that ultimately led to his being confined to a wheelchair. One of his doctors felt "bile" was the cause of his illness, and he was twice bled from the temples by leeches. There is another report that the first clinical description of MS may actually go back to the fourteenth century, to S. Lidwina of Schiedam, though this is disputed. The French called this new disease *sclérose en plaques,* the English *disseminated sclerosis,* and the Germans *multiple Sklerose.* All terms described the plaques scattered throughout the brain and spinal cord. The word *sclerosis* is from the Greek. After Charcot described the disease, MS became one of the more common reasons for admissions to the neurological wards, representing a new recognition of the disease, not an increased frequency.

But why did the astronomer get the disease? What triggered it in her body, or in anyone's body, for that matter?

A disease remains a mystery until the trigger is found. Last summer I got a call from my sister in Denver. She was having dizziness and visual problems. Her doctor had raised the possibility of MS and wanted her to get an MRI scan. It would be the irony of ironies, my sister coming down with MS, but of course it could happen; on more than one occasion I was certain that I was coming down with MS myself. As I listened to her story, though, there was something about it that didn't sound like MS. Her attacks were too violent; the visual symptoms didn't sound like the visual symptoms of an MS patient. In my mind, I compared my sister's words with the tens of thousands of words that I had heard over the past three decades from MS patients. They didn't match. I told her to hold off the MRI until I was in Denver and could examine her.

My sister and I love to play golf, and when I arrived in Denver, we drove straight from the airport to the golf course. In the car she told me her story again, and at the golf course I could find nothing abnormal on the brief neurological exam I performed in the corner of the clubhouse. Then, just before she hit her ball on the third hole, she began to have symptoms.

"I'm getting dizzy," she said.

"Try to hit the ball," I said.

She swung at the ball and missed.

I immediately examined her eye movements. They were jerky, not unlike those of an MS patient. But as I followed her attack over the next half hour, during which we left the golf course and headed for home, it became clear that my sister's attack was not MS. It reminded me of migraines. Both my wife and I suffer from migraines, and I carry pills with me at all times. I gave my sister one of my migraine pills, and it helped her. What triggers migraines? In my wife's case it is chocolate and red wine. In my sister's case, we discovered, caffeine can trigger an attack.

The big question that I set out to answer and that all MS researchers have been trying to answer is this: what triggers MS?

I recently gave a talk on multiple sclerosis to a small group of staffers at the Boston Home, a long-term care facility for MS patients who have severe disability and often need help transferring from wheelchair to bed. The group of thirty-five people knew all too well the effects of the disease but little about the latest theory on the cause of the disease. I stated the basic premise of the disease and then asked a question.

"In patients with multiple sclerosis," I said, "there are white blood cells in the brain that have come out of the bloodstream and entered the brain. In you and me there are no such white blood cells in the brain. My question is, what triggers the white blood cells to enter the brain? What is attracting them there?"

No one spoke. I repeated the question, and the audience now realized that my talk would not proceed until somebody answered. This time I gave a clue: "White blood cells are part of the immune system," I said. "Think of what the immune system does. Now, why in the world should white blood cells be triggered to leave the bloodstream and enter the brain?"

Five people raised their hands. I called on a woman sitting in the front row wearing a red dress.

"An infection," she said proudly.

"Exactly," I answered. "The simplest and most obvious explanation for multiple sclerosis is that there is an infection in the brain, and white blood cells are going there to eradicate the infection. Such a process leads to damage both from the infection in the brain and from the white blood cells traveling there." In an instant, it was clear to everyone that the key to MS was to find the virus that caused it.

This line of thinking, of course, has not been lost on scientists, and for a half century scientists have attempted to find a virus or other infectious agent that causes multiple sclerosis. Pierre Marie speculated in 1884 that MS was a complication of a large number of infections, including typhoid, smallpox, measles, whooping cough, and dysentery. The list of viruses and infectious agents that have been reported to cause MS is more than twenty. One of the first reports of a virus in MS came from Russia in 1946, when a new virus was reported to be isolated and transmitted from multiple sclerosis patients. However, the virus turned out to be rabies and in no way related to multiple sclerosis.

It is not always easy to find a virus that causes disease, and there have been instances of strange viruses or infectious agents causing disease in people. One of the most striking examples is that of an infectious agent that causes kuru, a fatal degenerative brain disease, and is the same type of infectious agent that causes mad cow disease. The infectious agent that causes kuru and mad cow disease is not a classic virus, and two Nobel Prizes have been awarded for research on this unique infectious agent. One went to Carleton Gajdusek in 1976 when he demonstrated that a disease unique to the Fore people in New Guinea could be transmitted by injecting brain matter into chimpanzees, and the reason these people came down with the disease was through practices of spreading brain matter on their skin or eating brain from people who had died. Then in 1997 Stanley Prusiner won a Nobel Prize for showing that this infectious agent was not a classic virus but was a replicating protein called a prion. Thus, if MS wasn't caused by an isolatable virus, perhaps it was caused by a yet-to-be-identified pathogen. Anything is possible in MS. No one would have predicted that there could be an infectious agent that was a protein, because proteins don't normally replicate, but two Nobel Prizes were given for discovering the agent and characterizing it.

If a disease is infectious, identifying the infectious agent is crucial to understanding, treating, and curing the disease. Two examples are AIDS and the HIV virus, and stomach ulcers and a bacterium called *Helicobacter pylori*. When the syndrome now known as AIDS was first described, no one knew what caused it. It was known, however, that the syndrome was seen more often in homosexual males and was associated with suppression of the immune system. Given that some homosexual male activity involved introduction of sperm into the rectum, some scientists postulated that this was what caused suppression of the immune system and AIDS. Experiments were even performed to test whether the introduction of sperm into the rectum of mice suppressed the immune system. That hypothesis and experiment are ludicrous to us now, but at the time HIV was a total unknown, and scientists tried to explore every possibility.

In the other classic example of an infectious agent unexpectedly discovered to cause a disease, for years the study of stomach ulcers focused on too much acid in the stomach and stress in people with type A personalities. But a physician in Australia named Barry Marshall postulated that stomach ulcers were due to an infection with a type of bacterium. Marshall was not taken seriously and ultimately proved his point by swallowing the bacterium himself, developing an ulcer, and then curing himself with a regimen of antibiotics and bismuth. Now the role of this bacterium, *Helicobacter pylori,* in ulcers of the stomach is well accepted, and those who had vociferously ridiculed Marshall simply shrug.

So what about MS?

Although, one of the promising areas of research in MS at the time I finished my training was to search for the "MS virus," because of Arnason's influence I thought the answer to MS lay more in the immune system than in the study of virology. Thus my wife and I took our two sons, then ages six and three, and headed back to Denver, where I would begin formal immunology training in the laboratories of Henry Claman and John Moorhead at the University of Colorado. I had a stipend of $12,500 from the Colorado MS Society and on weekends I moonlighted at St. Anthony's hospital on the west side of town where I grew up. I covered the intensive care unit and met helicopters landing on the rooftop with critically ill patients.

Henry Claman had made a major discovery in immunology, showing that the immune cells of the body, the lymphocytes, were of two classes, T cells and B cells. As we discussed earlier, T cells got their name because they come from the thymus; they are the cells that cause EAE. B cells make antibodies and are responsible for the elevated gamma globulin in the spinal fluid. Claman discovered that B cells couldn't make antibodies unless they received help from T cells. This was a huge discovery in immunology, like discovering that flowers can't grow unless they have water. In addition to having their own function in fighting off infection, T cells regulate the immune system. They provide help for other cells to work, and they are suppressors or regulators of the immune response. Over the course of my thirty years in medicine, our understanding of the complexity of different types of T cells would grow exponentially. And the story of T cells would intersect with that of MS in a crucial way. Today we believe that MS is caused by abnormal T cells that attack the myelin sheath (point one of the twenty-one points), and treatments that help MS may work by inducing other regulatory T cells to shut down the T cells that are causing the disease: T cells controlling T cells. But back then, the opinions in Claman's lab about what causes MS were just as divided as they were in the wider research community.

In order to study the immunology of MS, one had to know the fundamentals. Thus the immunology research I did in Claman's lab was basic and didn't even relate to T cells or MS, even though I would study T cells in MS for the rest of my life. I studied how B cells were triggered by immunoglobulin surface receptors, immersed myself in immunology journals, and attended basic immunology conferences. In attending conferences, I learned another of my first lessons about science: there's a big difference between reading someone's papers and hearing them present their work or discussing their research with them personally. Later in my career, like so many others, I would travel extensively to keep up with science.

Trying to find a cure for MS takes place in the laboratory as much as in the hospital. However, as I soon discovered, the ambiance of a basic science laboratory is different from that of the hospital. On the surface, the lab seems more relaxed than the hospital, where the pressure of illness and dealing with people creates an urgency. But the lab has its own urgency. Once an experiment starts, one is locked into a time frame for completing it, and that means coming in on weekends and staying late at night. Furthermore, many experiments don't work and have to be repeated. Without a positive reproducible experiment there is nothing to publish. If there is nothing to publish, there is no advancement and no money. In the lab it is truly publish or perish. In the hospital, doctors don't lose their jobs if they can't cure multiple sclerosis. Nonetheless, there is an advantage researchers have studying basic questions in the laboratory. The systems being studied are ones that can be manipulated and dissected in the search for clear answers. It is not like studying a disease such as MS, where one is more limited by the questions one can ask.

One day, in one of the impromptu meetings that so often happen in the hallways of scientific laboratories and are central to the exchange of ideas, Claman happened to catch myself and two Ph.D. colleagues, Steve Miller and John Moorhead, chatting. I was eating a sandwich that I had left in the hall to nibble on between experiments, Miller was drinking one of his many cans of Coke, and Moorhead was smoking his pipe.

This little colloquium led to an analysis of the ways in which the research community would have to solve the virus-versus-immune-system mystery that I have always found enormously helpful in conveying the main thrust of MS research during all these years.

"Well, gentlemen, what do you think causes MS?" Claman asked. That has been the opening sentence of countless discussions among scientists over the years as we confronted the origins of the monster.

"My working hypothesis is that it is an autoimmune disease," I said. "No one has ever been able to isolate a virus."

"What's the evidence that it's autoimmune?" Moorhead asked.

"Just because no one has been able to isolate a virus doesn't mean there isn't one there," Miller added.

"Gentlemen," Claman said, "at our next lab meeting, prepare two lists. On one write down what is required to prove that a virus or unique infectious agent causes MS. And on the second, do the same for proving that MS is an autoimmune disease."

I welcomed the exercise. We prepared the lists, and after the lab meeting, I wrote them down in one of my notebooks and referred to them so frequently that I had soon committed the lists to memory.

For the virus, there were five points (based on the postulates of a famous microbiologist named Robert Koch). First, the MS virus must be isolated from the majority of MS patients tested, and it should be found more commonly in MS patients who are sick than those who are well. Second, there should be antibodies or an immune response in the bloodstream of MS patients against the virus. Third, one should be able to reproduce MS in susceptible animals by infecting them with the MS virus. Fourth, no other virus should show the same effect as the MS virus. And fifth, vaccination to prevent the infection or treatment that eradicated the MS virus should cure MS or stop the disease process.

"Many viruses have been felt to be the MS virus," I said after the list was put on the board and I brought up the 1946 report that the rabies virus was the cause of MS.

"Absence of proof is not proof of absence," Claman said, echoing what I had discovered several years earlier in my experiments with MBP and EAE.

As Moorhead put the list for MS being an autoimmune disease on the board, I realized how similar the concepts were. Instead of a virus, the immune system was causing the damage, and the mechanism had to be identified. The autoimmune hypothesis also had five points. First, the immune response (T cells or antibodies) should be isolated from the majority of MS patients tested and should be found more commonly in MS patients who are sick than in those who are well. Second, the disease-causing immune cells or antibodies should react with something specific in the brain that they are attacking. Third, one should be able to reproduce MS in susceptible animals by creating disease-causing immune cells or antibodies or by transferring them from MS patients to animals. Fourth, no other disease-causing immune cells or antibodies should show the same effect as the MS autoimmune response. And fifth, treatment to destroy or inactivate the disease-causing immune cells or antibodies should cure MS or stop the disease process.

The lists were good ones, even though we all knew there could be exceptions to the rules. For example, an infectious agent (like the one that causes mad cow disease) may not evoke an immune response, and it may not be technically possible to transfer cells from people to animals because of species barriers. Nonetheless, pursuing the questions outlined on these lists has preoccupied most of MS research during the past quarter century.

After that encounter in the hallway, I continued to pursue my basic training in immunology. Then in 1976, a year and a half into my fellowship, I received an urgent call from Claman's secretary. "Howard, come to the main office," she said urgently. "There's been a big breakthrough in MS, and TV cameras are on the way to the lab. You know the most about MS, so Henry wants you to talk to them."

Claman told me that an article had just been published in the British journal *The Lancet* stating that a virus had been found to be the cause of MS. When I asked if that meant the autoimmune theory was dead, he said we'd have to see the data first. A reporter and cameraman arrived and told me they needed pictures of MS research and a statement about the virus that had been found in MS, though they didn't have a copy of the article for me to read. So I had a technician fill some test tubes with colored water for their pictures and gave the reporter the following statement: "One of the major theories about the cause of MS is that it is caused by a virus. For many years, scientists have been searching for the MS virus, and if one is found, it would be a major advance."

This was my first exposure to the media and its reporting about MS, and I would have many more such encounters over the course of my career. I discovered that interacting with the press is both problematic and necessary. I am always frustrated by the need to give sound bites, to try to explain complicated biology in a few sentences. But it is a chance to educate, to explain. In this instance, I was actually thankful that television was interested only in sound bites and that there were no follow-up questions. It wasn't until later in the week that I was actually able to get my hands on the article. Today, I could have gotten the article on the Internet. In 1976 there weren't even computers in most laboratories.

What did the article say?

The *Lancet* piece, published on February 28, 1976, was a major editorial whose title was "A Milestone in Multiple Sclerosis." It described work that was first reported in 1972 and which, as I soon found out, had evidently been reproduced by another laboratory. The editorial said, "A remarkable paper was published from the Institute of Research on Mental Retardation in which Dr. R. A. Carp and his colleagues claim to have demonstrated a transmissible factor, which appears to be virus, by inoculation of inbred mice with materials obtained from nine cases of MS. Their approach was disarmingly simple." It concluded, "These findings seem to remove multiple sclerosis from the group of diseases of unknown etiology or cause and place it squarely in the sector of the infectious diseases, although the precise nature of the virus is yet to be determined."

What did Carp and his colleagues do, and what happened between 1972 and 1976 that led to the *Lancet* editorial? Carp and his colleagues used a simple method to search for a transmissible infectious agent in MS. They took tissue samples from multiple sclerosis patients—three brains, a spleen, three blood samples and two samples of spinal fluid. They made an extract of the brain and the spleen and injected the extract into mice, either into the belly or directly into the brain itself. They found that within sixteen to forty-eight hours after the inoculation of the MS materials into the mice the white blood counts of the mice changed. They observed a decrease in the number of one type of white blood cell (neutrophils) and an increase in another subset (lymphocytes). As discussed earlier, neutrophils are white blood cells that fight off bacterial infections, while lymphocytes are specific immune cells. Whatever the case, it was logical that if there was an infectious agent in MS patients, it might have effects on white blood cells. Remarkably, the effects they observed persisted in the animals for almost a year. During this time, the mice were otherwise normal. This made sense in terms of MS patients, who at times appeared normal and theoretically could still be harboring a virus. The researchers then showed that they could transmit the effect by taking blood from the mice that had been injected with the MS material and injecting other mice with it. They did not observe the effect in mice injected with brain extracts from people without multiple sclerosis or with other diseases. Thus they appeared to satisfy three of the five criteria for MS being caused by a virus.

What was most striking about Carp's findings was that when they filtered the extracts that caused the effect on mouse white blood cells, they found that the transmissible factor was between 25 and 50 nanometers in diameter and was present at very high concentrations, 600 billion infectious units per gram of brain tissue. If this indeed was a virus, it was obviously a very unusual one because of its properties and its effects on white blood cells of mice. Nonetheless, there certainly could be an unusual virus causing MS; remember two Nobel Prizes were given for an infectious agent that caused mad cow disease and was like none other that had been seen before.

The essence of science is reproducibility. This is what makes science so different from other disciplines—say, the arts, the essence of which is to create a unique vision. No one immediately tries to exactly reproduce Picasso's paintings or Beethoven's music. But as soon as a new scientific truth is discovered, other scientists rush to exactly reproduce the findings. Why? All scientists build on the truths discovered by others, and they must be sure that the truths are real. Newton could see as far as he could, he said, because he stood on the shoulders of giants. But he needed to be certain he was standing on the right shoulders. If there was a virus involved in multiple sclerosis, entire laboratories would change the course of their investigation of MS and follow the new lead. But the new virus had to be confirmed.

Early attempts to substantiate the claims of Carp and his colleagues were not encouraging. In 1974 a paper from the National Institute of Neurological Diseases and Stroke in Bethesda reported no decrease in the white blood count of mice injected with material from MS patients. It was also known that white blood counts in mice are notoriously variable and are strain-dependent. To expand on their finding, Carp and his colleagues realized that it was too cumbersome to study their infectious agent by injecting it into animals and then waiting to measure white blood counts of the animals. They needed an in vitro (test tube) test for their agent. The same problem plagued the identification of the Kuru agent that could only be transferred from brain to brain of mice until a test-tube method was developed. Thus, two years later Carp and his associates created a special cell line, called PAM cells, from mouse cells, and they found that after infecting these cells with MS patient material, an infectious agent could be isolated. As before, these effects were obtained in 50 nanometer filtrates but not 25 nanometer filtrates. Again, positive results were claimed in eleven MS materials.

The event that led to the 1976 editorial in *The Lancet* was the publication in 1976 of two papers by an independent group headed by Dr. G. Henle in Philadelphia. They repeated Carp's work and observed it not only in mice but in rats, hamsters, and guinea pigs. They also confirmed that the agent was between 25 and 50 nanometers in size. In other words, the agent they had found appeared to be a virus somewhere between the size of the poliovirus and the influenza virus.

They also made a new observation, that the biological effect could be neutralized by blood from MS patients, and that the neutralization property appeared in the immunoglobulin fraction, the part of the blood that contained antibodies. Blood from normal Americans failed to neutralize the agent, but serum from relatives of MS patients gave positive results. Even more remarkable was that a strong neutralizing activity was seen in samples of blood collected in East Africa, where the disease was virtually unknown. Thus, they also had satisfied some of the major points we had listed during our journal club for MS to be caused by a virus.

It is easy to understand the scientific enthusiasm when this article appeared. The infectious agent became known as the multiple sclerosis associated agent, or MSAA. *The Lancet* wrote, "The advance of the

MSAA offers the hope that while the control of multiple sclerosis may still be some distance away the pace of research is quickening."

However, one and a half years later, on October 15, 1977, a small letter to the editor appeared in *The Lancet*. The letter was five paragraphs long and was simply entitled "Multiple Sclerosis Associated Agent." It was signed by Henle and colleagues and by Carp and his colleagues. In the first paragraph they summarized their findings with the MSAA, its size, its transmissibility, and its unusual properties. In the second paragraph they stated that their observations had not been originally confirmed but were later confirmed. In the third paragraph they explained the search for MSAA neutralizing antibodies and the potential relationship between MSAA and another infection found in East Africa that was related to Epstein-Barr virus (which causes mononucleosis). In the next paragraph, however, they wrote that they were bothered by the *Lancet* editorial one and a half years earlier—that the unwarranted publicity had raised false hopes and brought numerous letters from patients in many parts of the world. Then they wrote that research on MSAA should be discontinued. What had happened?

As we've discussed, the essence of science is reproducibility. Under the auspices of the National Multiple Sclerosis Society a large number of new specimens were provided to Carp to be tested for the presence of MSAA. Unfortunately, after extensive testing, Carp and Henle's results could not be confirmed with these new samples. Carp and Henle gave no explanation for the inability to reproduce their findings and wrote that the validity of their observations could only be judged at a future time. Since then, there has been no further study of the MSAA.

The remarkable MSAA story, with its unfortunate ending, exemplifies the highly charged nature of research into a disease such as multiple sclerosis. In our own laboratory we have seen findings that at the time we felt were real but which later turned out to be artifacts (results of some extraneous factor, not related to the hypothesis being investigated). Often the artifacts are discovered when we try to do the experiment in a different way, only to discover that it doesn't work. What we thought we were measuring was really something else.

In August 1999 Steve Miller and I attended an international

workshop in Brighton, England, organized jointly by the Multiple Sclerosis Societies of England and the United States on the role of infections in multiple sclerosis. The workshop brought researchers together from around the world to critically review the status of viruses and infectious agents in MS. During one of the breaks from the conference, we found a nearby park where we could walk and discuss science, a routine we developed over the years when we were at the same conference. Steve was studying a virus called Theiler's virus, which causes an MS-like disease in mice. We reflected on the role of viruses in MS and concluded that no one had yet found an MS virus. Between the time we were fellows together in Henry Claman's lab and that day in Brighton, there had been many other reports of possible MS viruses (parainfluenza virus, MS-associated retrovirus, coronavirus, SMON-like virus, herpes simplex virus, canine distemper virus, HTLV-I virus), but none held up under scrutiny as the MS virus. The latest infectious agents reported to be associated with MS were a herpes virus called HHV-6 and a type of bacterium called *Chlamydia pneumoniae.* There was controversy at the conference about whether these agents were the direct cause of MS.

I used the opportunity to ask Richard Johnson, a professor of neurology at Johns Hopkins and one of the world's experts on viral infections of the nervous system, both about the Carp agent and why it has been so hard to find the MS virus.

"First of all," he said, "there may not be an MS virus. If there is, there is something unique about it, and because it is not easily identified, investigators end up reporting lab contamination or the discovery of viruses that are present in all individuals but are not specific to MS."

"What went wrong with the Carp agent?" I asked.

"No one really knows," he said. "It most probably relates to unknown variables that have plagued all reports of viruses in MS. There are even rumors that an overzealous technician in one of the labs may have contributed to part of the positive results."

At the end of the Brighton conference, most agreed that to date no unique MS virus or other infectious agent had been identified. Nonetheless, there was ample evidence that viruses or infectious agents were involved in MS by somehow triggering the immune system to attack the brain—like the red wine and chocolate that trigger migraine attacks in my wife.

Increased MS attacks are associated with viral infections, especially upper respiratory tract infections, even though MS patients tend to have fewer viral infections than people without MS, perhaps because their immune system is activated. Nonetheless, there is no evidence that flu shots trigger MS attacks. Some believe that common viruses may trigger MS in susceptible individuals, such as Epstein-Barr virus. In fact, MS patients have increased antibodies in the spinal fluid against many viruses.

There have been reports of MS epidemics, the most intensely studied occurring in the Faroe Islands located northwest of Scotland, where MS presumably did not exist until the British army arrived during World War II. Some argue that they brought an infectious agent, perhaps with their dogs. Others feel the epidemic was simply better recognition of the disease. There is no increased incidence of MS in spouses of those with MS, something one might expect if there was an easily transmissible infectious agent.

MS is more common in certain parts of the world and is less common the closer one is to the equator. It appears to have originated in northern Europeans and then spread. There is a high incidence of MS in northern Europeans, U.S. Caucasians, and Canadians. MS is rare in American Indians, African blacks, and Eskimos. The reason for the geographical distribution is unknown, but probably relates to several factors, including genetics, incidence of viral or other infections, and even the amount of sunlight. Some studies suggest that where one spends their childhood may affect the risk of contracting MS; perhaps exposure to something in the environment at an early age makes one more susceptible to contracting MS as an adult. MS occurs in children, but it is rare before puberty. Thus, it appears that viruses or other infectious agents may trigger or exacerbate MS by affecting the immune system in genetically susceptible individuals.

MS is more common in women. Approximately two-thirds of those with MS are women. The incidence of other autoimmune diseases such as rheumatoid arthritis is also higher in women and most probably relates to hormonal factors that influence the immune response.

Certainly the story of the origins of MS has turned out to be more complex than I or my fellow researchers had first imagined. But as we will see in the following chapters, we have in fact made remarkable progress in tracking down one after another important clue in the complex MS equation with which to solve the riddle. And although the hopes that a simple viral cause for MS could be found and a vaccination readily developed suffered a setback, the research community was pressing forward with several other promising ideas related to immune-based therapy. I soon became involved in an exciting clinical experiment to test one of these.

"The origins of MS are so complex," the astronomer's husband once said at the end of one of her visits to my office.

"Not as complex as the universe," his wife said.

6

PLASMA, PLACEBOS, AND CLINICAL TRIALS

GLORIA EDWARDS IS an MS patient who stands out vividly in my mind. This feisty woman, who could no longer work in her family's bakery because of her MS, was the first patient who agreed to participate in an experimental trial I initiated. She exemplifies the courage of those who have volunteered to undergo a treatment never before tried in MS. She also is a symbol of the hope all MS patients have that something new will be found that will help them. Initiating my first trial was a very exciting moment for me. I had read extensively on MS, I had studied immunology, and I had experimented on hundreds of mice in the laboratory. Now I would be initiating a clinical trial that, if successful, had the potential to actually help MS patients. Ultimately that was what I wanted to accomplish—that was my dream.

AFTER MY TWO YEARS in Henry Claman's lab, I moved back to Boston and joined the laboratory of Bernie Fields, in the department of microbiology and molecular genetics at Harvard. With Bernie, I would be studying viruses, because even though the Carp agent had turned out not to be the MS virus, everyone still believed that viruses were somehow involved in the cause of MS.

I had a small desk in a shared office in Building D of the Harvard Medical School, the building with the well-worn saying etched onto the outside wall, "Life is short, the art long, the occasion instant, experiment perilous, decision difficult." There I sat, having decided to devote my life to the long art.

In the three years I was in Bernie's lab, I would study reoviruses, which don't cause any human disease. Rather, we study these animal and plant viruses as models to learn the basic mechanisms of how a virus interacts with the brain and with the immune system. The reovirus is a great model because it has only ten genes that can be relatively easily manipulated. During my time in the lab, by mixing genes between strains of reovirus, I carried out a now classic series of experiments in which I defined one of the basic mechanisms by which a virus causes disease and interacts with the immune system. I discovered that reoviruses use a specific structure on their surface to infect brain cells and to stimulate the immune system. Based on these experiments, I was able to move up a major rung on the academic ladder, become an assistant professor, and establish my own laboratory. (The principle I discovered also explains how HIV, the virus that causes AIDS, infects cells and why some people are resistant to it—because their white blood cells don't have a structure on their surface to which the virus can attach.)

By studying reoviruses, I learned that the core of medical and biological research is discovering the key variables or features that are responsible for a biological observation. It is difficult, because in biology there is rarely only one variable. Why do some people get MS and others don't? Why in some people is MS mild and in others it is severe, and why does it take different forms? Reoviruses are easy compared to MS.

Even though I was working on viruses unrelated to MS, I could not stop wondering about the cause of the disease. I had decided not to see MS patients while getting basic immunology training in Claman's lab, but I was now ready to study the disease itself, and for me the study of MS had to involve caring for MS patients and trying to find new treatments for MS. I believed one could not truly study a human disease without directly confronting those afflicted with the illness. So I began to see MS patients at what was then the Peter Bent Brigham Hospital.

Over a period of months I kept reviewing in my mind the theories of MS and the possibilities of treatment. I thought of the ventricles. I thought of the areas around the ventricles, where it looked as though a toxic substance had leaked out and caused damage. If MS was an autoimmune disease, then one had to identify which part of the immune system was attacking the brain. I had learned that the immune system has two components that could be causing the damage associated with MS: cells and antibodies.

I began to consider the possibility of antibodies causing the damage of MS, and reviewed the five requirements we had discussed in Henry Claman's lab to prove the autoimmune theory of MS. The first two points would require that antibodies against myelin exist in MS and that they be found at higher concentrations in MS sufferers than in people without MS. Reviewing the literature, I found studies suggesting that antimyelin antibodies existed in the blood of MS patients, but they were not conclusive.

The third requirement was the ability to reproduce MS in animals by transferring antibodies from MS patients into animals. I had just read an article in the *New England Journal of Medicine* reporting that this had indeed been shown for another neurological disease, myasthenia gravis, which causes muscle weakness. Myasthenia gravis is caused by antibodies that interfere with a chemical called acetylcholine, which is released by nerve fibers to stimulate the muscles. Patients with myasthenia gravis have antibodies against the acetylcholine receptor, and the *New England Journal of Medicine* article showed that if one purified antibodies from the blood of patients with myasthenia gravis and transferred them into mice, the mice came down with a disease that looked like myasthenia gravis.

It was the fifth and most crucial of the requirements that I thought about most: the requirement to show that giving a treatment to the MS patient that destroyed or inactivated the disease-causing antibodies should cure MS or stop the disease process.

Shuttling between the Brigham and Bernie's lab one day, I thought that if MS patients had abnormal antibodies directed against the myelin that were indeed playing a role in the disease, shouldn't MS patients improve if we removed the antibodies? There is a way to remove antibodies from a patient's system, called plasma exchange. The patient is hooked up to a machine, blood is drawn out from one arm, and the antibodies are filtered out of the blood with the plasma. Then the filtered blood is returned via a needle in the other arm. I was quite excited by the idea and called Dave Dawson. Even though there was no formal proof that antimyelin antibodies existed in MS, would he be willing to carry out a study in which we took MS patients who were not doing well and had not responded to corticosteroids and treat them with plasma exchange?

Dawson said yes. There was a good rationale for the experiment, and plasma exchange was being performed on patients from the Brigham at the Children's Hospital blood bank across the street.

This was a major step for me. I was about ready to confront MS head on. I would be doing an experiment on MS patients that was potentially a form of treatment. In the lab, I had learned how to culture lymphocytes and grow viruses; now I would begin to learn about conducting clinical trials. In the lab, much of what we did was several steps removed from treating MS. In a clinical trial, anything that was observed from the treatment had to be directly related to MS even if we didn't understand it, because it occurred in a person with the disease process.

As we began discussing the clinical trial, I immediately realized that there were many new lessons to learn. The biggest lesson was that one could not control the variables in a clinical trial in the elegant way I had shuffled genes between different strains of reovirus in my studies of mice. The mice we studied were genetically identical, whereas each person with the disease is different, a unique individual. Furthermore, depending on the treatment being tested, clinical trials go through stages. Usually there is a limited preliminary study to make sure the treatment is not dangerous and to look for positive effects. Only then can one move to larger studies and actually prove that the treatment works. I also realized clinical trials take much more time than lab tests, and one doesn't publish a lot of papers. Yet in order to get promoted in academia, one has to publish papers. Even though promotion committees say it's the quality of one's work, not the quantity, that is important, everyone knows that in fact committees count numbers of papers published. In addition, clinical trials often involve large teams, and it's hard to get individual credit. In 1983 in Buffalo, at the first workshop devoted solely to clinical trials sponsored by the National Multiple Sclerosis Society, Barry Arnason said to me, "You can't do clinical trials and get promoted." Times have changed. NIH and the National Multiple Sclerosis Society have now established training programs specifically for doctors who want to do clinical trials, and Harvard Medical School has established a clinically based academic track. Nonetheless, even with these changes, getting ahead while doing clinical work still isn't easy. I realized that I was lucky that I had already been publishing papers on my experimental work with reovirus. That work gave me the luxury to embark on a dual path: performing clinical trials in patients with MS without sacrificing academic advancement. All I needed was the energy to do both.

I held long discussions with Dawson and the people at the blood bank at Children's Hospital, where the plasma exchange would be done. It was relatively easy to interest them in the study, because MS was a crippling disease with no effective treatment. There were reports in the literature not only of antibodies against myelin components in MS patients but also other serum factors that might play a role. Thus we could benefit patients by removing more than one potentially toxic factor from the plasma. The first question, of course, was which patients to treat. We elected to treat patients who were not in the midst of an acute attack, as it would be difficult to distinguish whether the effect we observed was related to the plasma exchange or to natural recovery from the attack. Thus we chose patients who had had progression of their MS for at least a year despite treatment with ACTH or steroids.

The next question was whether we should do the plasma exchange alone or combine it with other treatments. There was evidence from the use of plasma exchange in other diseases that for plasma exchange to work, one had to give steroids or other immunosuppressive therapy to keep the antibody levels from going right back up after the exchange. There were no reports of plasma exchange in MS, and it was crucial to get a positive effect. We didn't want to perform plasma exchange, not have it work, and then be told that we didn't give a strong enough regimen because we didn't give immunosuppressants at the time of our plasma exchange. So we decided to treat patients with steroids and an oral immunosuppressant called azathiorprine, and administer three to five plasma exchanges. We would perform a spinal tap before the first plasma exchange and after the last one to measure whether we had lowered the elevated levels of gamma globulin in the spinal fluid, even though we didn't know what the elevated gamma globulin was reacting against. Before we could start, we had to have approval of the hospital's human studies committee, which took three months from the time I submitted the protocol.

Finally we were ready to begin. The first patient, Gloria Edwards, was admitted on Sunday afternoon to the hospital's clinical research center on a floor called E-Main, just off the hospital pike. She was forty-two years old, a tall thin woman, who was a patient of Dave Dawson. I came into the hospital to perform the neurologic evaluation, spinal tap, and to obtain informed consent. Gloria Edwards' symptoms had begun at age twenty-four, when she had episodes of double vision as well as loss of vision. She was well until five years later, when her disease became progressive. This led to weakness and spasticity of her legs and weakness of her right hand. She now could not get around without a walker. She was very excited about the study and had brought a cake from her family's bakery for the clinical research staff. I admired her courage and willingness to endure the large needles in both arms that would be required for the plasma exchange.

"How do you feel, Gloria?" I asked as I sat at her bedside with the chart resting on my knees.

"I'm very excited," she said. "I've been looking forward to these treatments for over a month, ever since Dr. Dawson told me about them."

I tried to explain that there were no guarantees that these treatments would help her, that it was an experimental protocol.

"I know," she said, "but you wouldn't be trying it if you didn't think it had a chance of helping, would you?"

I could not refute her simple logic and began to realize the complexities of helping patients to manage their expectations when they entered a clinical trial. How could a patient not hope that the treatment might help? I tried again to explain that the treatment was experimental.

She nodded, but it was obvious to me that in her mind she felt the treatment was going to help her. That's what she was focusing on, not that it was experimental. There was nothing I could do to change her mind, and on one level she was right.

I read over the consent form with her, explaining the risks, the fact that it was experimental, how the procedure would be done, and her option to withdraw from the trial at any time she wanted. She nodded as I spoke, then signed the consent form, even though it was difficult for her because of weakness in her right hand.

Before I left she asked a simple question: "Has this been done before in MS patients?"

I looked at her, at the consent form, and then at the walker next to her bed. "No," I said. "You're the first one."

I later learned that at the same time we were embarking on our study, Peter Dau in Chicago had also begun a study of plasma exchange in MS patients. In science and medicine, it is probably the rule rather than the exception that more than one group or individual are independently testing identical hypotheses at the same time. It was true for the discovery of the double-helix structure of DNA; it was even true for Darwin and the theory of evolution. The race to be the first is important for the individuals involved and for the Nobel Prize committees, and sometimes there are unique insights—perhaps Einstein and the special theory of relativity is the prime example—but in the long view of science and medicine, it doesn't really matter who is first. I don't know whether Gloria Edwards was indeed the very first MS patient to receive plasma exchange, and in the end, it doesn't matter. But she certainly was the first in our hospital and one of the first in the world. In treating her, I stepped onto a playing field that I and other investigators would learn over the next quarter century was also a minefield.

I waited until after dinner to perform the spinal tap on Gloria Edwards so she could lie flat after the tap to decrease the risk of getting a headache, rest, and then go to sleep. As I inserted the needle into Gloria Edwards' back for the spinal tap, I thought back to the spinal tap I had performed on John Saccone seven years earlier to check for raised gamma globulin and establish a diagnosis. Now as part of my first clinical trial, we would see if plasma exchange affected the gamma globulin in the spinal fluid. Luckily, I had no trouble with the tap and sat with Gloria Edwards for a while afterward and answered last-minute questions she had.

Later, at the nurses' station, I leafed through her chart and thought about the theory that supported what we were about to do. I imagined the antibodies attaching to the myelin sheath and how the plasma exchange would be removing them. Whether removing them would actually have any effect on her illness was yet to be seen.

I appeared on the ward at seven-thirty the next morning and wheeled Gloria Edwards myself down the pike and across the bridge to the Children's Hospital blood bank for her exchange. As prescribed in the protocol, she would receive a total of three plasma exchanges, on Monday, Wednesday, and Friday of the same week. A large needle was placed in one arm to remove her blood. The blood was then filtered through a machine that removed the plasma but did not affect her white blood cells or red blood cells. Albumin was added to the filtered blood as a replacement for the proteins in the plasma, and the filtered blood was then returned through a large needle placed in her other arm. I stayed to watch the machine being hooked up and for the first thirty minutes of the plasma exchange, then I went back to the Brigham and finished my ward rounds.

The first exchange was over by noon, and I immediately checked on Gloria Edwards when she returned to the ward. "How do you feel?" I asked, trying to hide my excitement.

"A bit washed out," she said.

I performed a neurological exam and found no change.

When I stopped by on Tuesday morning, she no longer felt washed out.

"Do you feel any better?" I asked, which of course was a leading question.

"I think I'm a little better," she said, "but I'm not sure."

She underwent her second exchange on Wednesday and her third exchange on Friday. I continued to check on her each day, but it wasn't until Sunday, on the day that she was to leave the hospital that a clear difference was observed. When I came in to examine her, she had a big smile on her face.

"Dr. Weiner, I really feel better," she said. "My arms and legs seem stronger, and my balance is better." To get a more objective measure, I timed how long it took her to walk twenty-five feet with her walker. When I had measured her prior to treatment, it took ninety seconds. Now it only took forty seconds. She also was able to sit up straighter and could lift her arms above her head. There was little question in my mind that she was better. Indeed, after she left the hospital, she reported that she could swim the sidestroke with the scissors kick, something she had not been able to do for three years.

I was ecstatic with the first result, but as Dave Dawson reminded me, one had to wait to see what happened with the rest of the patients before we'd really know anything. We'd also have to see what happened to Gloria Edwards after a few months.

We treated a total of eight patients in our study. Each received three to five exchanges, and we found that six of the eight patients treated improved. One had improvement in his handwriting, and we published a handwriting sample from him before and after treatment. Two patients were unchanged. However, as we followed the patients, those who had improved returned to their pretreatment level within a few months, including Gloria Edwards. We gave her a repeat treatment six months later, and she improved again, although only temporarily.

We published our results in the October 1980 issue of the journal *Neurology* and entitled it "Plasma exchange in multiple sclerosis: a preliminary study." We concluded that although the role of plasma exchange in MS remained to be defined, we had observed no adverse side effects, and additional studies were needed. The lack of side effects was itself an important finding for an initial clinical trial. In the years that followed, when MS clinical trials became more commonplace, there were treatments that had to be stopped because of side effects.

The excitement of carrying out my first clinical trial in MS gave way to the harsh reality of realizing what we had and had not achieved. There was no way we could be sure that we had truly done something significant, as there was no comparison or control group, and even if the improvement was real, we had no idea what actually caused it or why it did not last long. There were too many variables: the plasma exchange, the steroids, and the azathiorprine. One variable we had eliminated was the possibility that the improvement was caused by a lowering of body temperature. We postulated in our paper that if the improvement was real, it could have been related to the removal of either antibodies or the other serum factors. We indeed found that treatment decreased the gamma globulin level in the spinal fluid in some patients, so our treatment had the biological effect we had hoped for. However, those patients with a lower gamma globulin level in the spinal fluid were not necessarily those who appeared to respond best to therapy.

A few months later I pored over the hospital charts on the patients we had treated in the experiment and reflected on the results. I reconsidered our major hypothesis, that MS was caused by antibodies that attacked the myelin sheath. There was still work to be done in the laboratory to identify and quantify any abnormal antibodies against myelin in MS patients.

Furthermore, although we had seen positive results in some of the patients, the more I thought about it, the more I was worried that what we had seen could have been a placebo effect, something not related to treatment at all. I will never forget an experience I had in medical school that involved a woman who had difficulty breathing that we thought was caused by blood clots moving to her lungs from her legs. With the technology available at that time, we couldn't be certain that there indeed were blood clots in her legs. A heated debate ensued, and it was decided that the only way we could test whether she had blood clots was to treat her with a blood thinner and observe her for three to four days to see whether her breathing improved. After a heated debate among the doctors, it was finally decided to give her therapy.

Four days later, as we stood at her bedside, she reported that she felt better and her breathing indeed had improved.

"I knew she had blood clots," one of the doctors said.

The doctor who had argued for not treating her shrugged, and we stepped into the hallway. We then had another discussion about whether the dose of the blood thinner should be decreased to reduce risks of bleeding. It was decided to cut the dose in half.

One of the doctors turned to the nurse and told her to decrease the blood thinner by half.

The nurse was puzzled. "We didn't know that she was supposed to be getting the blood thinner," she said. "We thought it was supposed to start next week."

The woman had never received any thinner.

How did we know that it was the plasma exchange regimen that made our patients better, even if it was only for a few months? How does one establish cause and effect in such a difficult disease as multiple sclerosis, or in any disease, for that matter? There are many instances in medicine of presumed cause and effect that did not stand up to scrutiny. A recent example is silicone breast implants, which were thought to cause autoimmune diseases (insurance companies paid out billions in claims), but on close inspection, the association could not be proven. The only way to show true cause and effect is through a carefully controlled clinical trial or laboratory test. Doctors like to believe they are helping patients and patients like to believe they are being helped, but both can be fooled.

THE FIRST CAREFULLY controlled clinical trials in multiple sclerosis involved the drug ACTH, and the hypothesis being tested was that steroids could help MS. Steroids were a wonder drug when they were first introduced and have helped patients with countless diseases. I remember a colleague once saying, "Don't let anyone die before you give them a shot of steroids."

If MS was an inflammatory disease of the brain and steroids were a strong anti-inflammatory medication, it was an obvious move to treat MS patients with steroids. In the 1950s there were several uncontrolled reports on the short- and long-term effects of ACTH on the clinical patterns of multiple sclerosis. ACTH stimulated the adrenal gland to make cortisone, and I had treated John Saccone with ACTH when I was a resident. Between 1958 and 1961, Miller and colleagues studied forty consecutive patients who had an MS attack as defined by new signs and symptoms in the past fourteen days. Half were randomly chosen to be hospitalized for three weeks and receive ACTH given intramuscularly; the others would receive a placebo (an injection of saline). In 1961, they reported positive results in the treated patients, especially in those who had optic neuritis, or eye inflammation. A controlled trial of fifty-five patients by Alexander and Cass in 1963 and of seventy-three patients by Rinne and colleagues in 1968 also suggested a benefit from ACTH.

While the Miller study was being performed, a conference was convened in the United States in 1960 to review different therapeutics agents for the treatment of multiple sclerosis. The conclusion: none were of established value. Nonetheless, because many physicians were using ACTH to treat MS and there was some suggestion of benefit, it was decided that ACTH was worthy of definitive clinical testing to determine whether if given over a short period of time it could help MS relapses. The first cooperative multicenter study for the treatment of multiple sclerosis began in 1965. The investigators wrote in their final report, published in 1970, that the purpose of the study was not only to test whether ACTH was better than a placebo but to determine whether a therapeutic agent could be reliably tested in MS.

Patients in ten U.S. academic centers were randomly chosen to receive ACTH or a placebo. The study was double-blind—neither patient nor physician knew what treatment was being given. Patients were hospitalized for two weeks and were evaluated twenty-four hours before treatment began and on weeks one, two, and five after treatment. Only patients with a clear attack within the previous eight weeks were eligible. Those who were having their first attack or who had complicated or advanced disease were not included. In addition, to control for people naturally improving from an attack or not having a bona fide attack, there could be minimal or no evidence of improvement at the start of treatment.

The investigators viewed it as a modern study and wrote, "The use of the computer provided the opportunity for extensive analysis of the great mass of data accumulated." The physicians used clinical rating scales as defined by Kurtzke and carefully recorded all the different parts of the nervous system examination. There were also many quantitative tests of neurological function, including speed of tapping the hand, how long the patient could stand with eyes open or closed, measurement of hand sensation, and the length of time the patient could feel vibration from a tuning fork.

What were the results? First, it was shown that the patients and physicians were indeed blinded and that the physicians could not predict which patient was on treatment versus placebo. However, although the degree of improvement in the patients treated with ACTH was statistically significant by some of the quantitative methods, at no time was the improvement particularly obvious or that different between the two groups. At five weeks, 69 percent of the patients treated with ACTH showed improvement, whereas 58 percent of those treated with placebo also showed improvement. An effort was made to obtain follow-up examinations and clinical histories, but it was not easy. Only 90 of the 197 patients were followed up, and there were not enough data to present follow-up results.

Thus, in this first multicenter cooperative trial of an MS treatment the doctors did not do substantially better in helping patients recover from an attack by treating with ACTH than did Mother Nature. It was difficult to show differences in recovery once the attack began. The confounding variables were both the ability of the body to heal itself and the placebo response. And though the study employed sophisticated clinical testing and used computers to analyze the data, the MRI technology that has since become crucial for testing MS drugs and helping them to get approved by the FDA had not yet been invented.

In their discussion of the results of the 1965 ACTH study, the authors emphasized that the placebo effect in a well-ordered, seriously applied therapeutic effort can be a powerful influence. They wrote that their observations should temper the enthusiasm of those who advocated a specific therapy for MS unless the trial is adequately and appropriately controlled. While the study confirmed the conclusion by Miller's group in 1961 that short-term, high-dose use of ACTH caused improvement, it could not be determined whether the long-term improvement was greater than that caused by placebo, as the patients were only followed for five weeks. There were no major complications, but no information was obtained to explain the mode of action or means by which the improvement may have occurred. Finally, although the results of the study provided no basis for enthusiastically recommending the use of ACTH in the treatment of multiple sclerosis, doctors continued to prescribe it. In retrospect, one of the major problems with the study is that it didn't address the central question in MS therapeutics, which was not whether one could treat individual attacks but whether it was possible to prevent attacks and the accumulation of disability over time. ACTH is no longer used today to treat MS attacks, having been replaced by the use of intravenous steroids.

The placebo effect is powerful, but it is real. Dale McFarlin, one of the major figures in the study of multiple sclerosis, is quoted as having said, "When I get MS, put me in a clinical trial and make sure I get a placebo." That quote reflects a cynicism that has plagued multiple sclerosis clinical trials over the years but is also part of the reality of studying the disease. The first book on the treatment of multiple sclerosis published by the National Multiple Sclerosis Society was not called *Therapy of Multiple Sclerosis,* but had the title *Therapeutic Claims.* The implication of the book's title was that there were "claims" of efficacy but not real efficacy and the MS Society book would show how treatments that were claimed to be efficacious indeed were not. Rather than a blueprint for how to treat the disease, the book was a cautionary note that treatments for the disease had not yet been found. There have been four editions of the *Therapeutic Claims* book until the title finally changed. In 2001, the fifth edition of *Therapeutic Claims* had a new title, *Multiple Sclerosis: The Guide to Treatment and Management.* It took four decades of clinical trials before that title could be changed.

In 1980 we confronted the same problem with our study of plasma exchange that investigators had faced with the study of ACTH in 1960. In addition to our study and that of Peter Dau, other investigators began to report positive results in small trials of plasma exchange. How to prove that there indeed was a clinical effect? A meeting was held at the National Multiple Sclerosis Society headquarters in New York, and it was decided to perform a large, multicenter trial to determine if there was any role for plasma exchange in the treatment of MS. We also decided to study patients whose disease was acutely worsening. A grant proposal was submitted to the National Institutes of Health, and it was approved after a series of revisions. This was much different from Dawson and myself treating eight patients at the clinical research center at the Brigham. There were investigator meetings, a safety monitoring committee, statisticians, and countless forms. A debate ensued about how to include a placebo group. The investigators felt that one couldn't get maximum effect of the plasma exchange unless immunosuppressive treatment was given to keep antibody levels down. So both groups received immunosuppressive treatment, and one group received real plasma exchange and one group received sham plasma exchange. For the sham exchange, blood was removed from one arm, passed through the machine without being filtered, and then returned to the other arm. (I myself underwent both a plasma exchange and a sham plasma exchange to experience what the MS patients were going through and to demonstrate that the treatment was safe. I didn't tell my wife until it was over.)

Large, multicenter cooperative trials take time. Our study was begun in 1984, but the results were not published until 1989. One hundred sixteen patients with acute exacerbations of disease were in the trial. They were treated over eight weeks and received eleven real exchanges or eleven sham exchanges, a much longer regimen than the three exchanges over one week Dawson and I had done. All patients received ACTH shots, as it was known that ACTH had some positive effect on attacks of MS and we didn't want patients in the midst of an acute attack to be without any treatment. An oral immunosuppressant, cyclophosphamide, was given to maximize the effect of the plasma exchange. The results were positive, but not dramatically so. There was increased clinical recovery from the attack at four weeks in patients whose disease course was primarily relapsing-remitting (as opposed to those with a progressive component to their disease), and patients recovered from their attacks quicker. Nonetheless, there were no long-term benefits. We did show, however, that the plasma exchange decreased blood levels of immunoglobulin and other blood components such as complement and fibrinogen, thus we had indeed "cleaned out the blood." Given the lack of long-term benefit and an only modest effect on clinical recovery, our study showed that adding plasma exchange to ACTH did not offer a great deal to patients. Dr. Khatri in Milwaukee later carried out studies of plasma exchange given more frequently, but again without major success.

In 1987, while we were finishing up our study of plasma exchange, Milligan and colleagues published a paper on the use of a different type of steroid, intravenous methylprednisolone, for the treatment of attacks of MS. It was a double-blind, placebo-controlled trial that involved twenty-two patients in the midst of acute relapse and twenty-eight with the chronic progressive form of the disease. They reported clear effects in patients in relapse but not as much of an effect in those in the progressive stage, just as we had observed in our plasma exchange trial. Other investigators have found similar effects, especially in patients with optic neuritis. Thus, the use of short courses of IV methylprednisolone (three to five days) for acute attacks has become common practice in the treatment of MS and is the extension of the early studies of ACTH. IV steroids help people recover faster and in some instances can halt an attack, although there is no proof they have a major impact on long-term disability or disease activity. Trials using intermittent IV steroids are currently under way to determine if there are effects on disability.

What is the status of plasma exchange today? In 1999 Weinshenker and colleagues at the Mayo Clinic demonstrated that plasma exchange may benefit the rare MS patient with a severe MS exacerbation that does not respond to IV methylprednisolone. His results were consistent with what we had found in our multicenter trial, and plasma exchange or strong immunosuppression is now used in this small patient subgroup by some physicians. Furthermore, as we have learned more about subgroups of MS, we think there may be some patients—those with high levels of antimyelin antibodies—who may benefit from plasma exchange. Nonetheless, plasma exchange turned out to make barely a dent in the armor of the monster.

Each new therapeutic agent that has been tested in multiple sclerosis, from the time of the first ACTH trial onward, has faced the same hurdles I've described here: how to prove the drug was actually working, how to study patients over long enough periods of time to measure a clinically important effect, and how to move from observations on small numbers of patients to generalizations about large numbers of patients. From the time a drug is first tested until it is ready for widespread use takes years. We now have drugs that have been shown to help MS in large-scale trials, although, like the first ACTH trial, many trials have failed.

Although some drugs have shown positive effects in large trials, we still must address the challenge of finding therapies that work in individual patients. MS can have an unpredictable course, with spontaneous remissions, and available treatments are not 100 percent effective. This makes it difficult both for the patient and for the doctor, who must decide together on an individual course of treatment.

Two of my patients help illustrate the conundrum. Arthur Strauss is an investment banker whose MS began with limping in one leg. It has slowly progressed to where he now uses a cane. However, his disease does not change rapidly, and sometimes he feels he is doing well. After he began a new treatment, I asked him a year later how he was doing. "Well, I think I'm doing fine," he said, "and things have been relatively stable, though I may have worsened slightly. To tell you the truth, I don't know how I would have done if I wasn't taking the medication."

Another patient of mine, Sylvia Pender, a mother with two young girls, has the relapsing-remitting form of the disease. She is on treatment, but once a year, often in the spring or fall, she will have a minor attack. Her attacks take different forms. Sometimes it is numbness in the leg; other times it is problems with her vision. Other times the attack can simply be fatigue. The attacks last anywhere from one to three weeks. She generally recovers from the attacks, and if she has difficulty walking, I treat her with steroids. Each time I see her, she says she is doing fine, but if one carefully examines the changes in her condition over the past five years, it's clear that she is not able to function as well as she did in the past.

I sat with her and her five-year-old daughter, whom I had given small tongue depressors to play with, and we discussed her illness. She was in the midst of another attack, with weakness in one of her legs.

"Should we give you a treatment to help you recover from the attack?" I asked.

"What would it be?" she said.

"Steroids given by infusion," I said.

"I've taken the steroids before," she said, "and I'm not sure they make that much of a difference."

"I know," I said. "But if we wait until after your attack is over and you don't do well, it will be too late to give you a treatment for the attack."

She looked at me, then at her daughter, but didn't answer. I myself was frustrated at the idea of giving a treatment and not being sure that it indeed would make a difference in an individual patient. (Actually, there is a clinical trial design in which an individual patient first receives treatment or placebo and then is "crossed over" to receive the other, but it is imperfect at best, as treatment and placebo effects can influence each other.)

I thought for a second and finally said to her, "You know, Sylvia, the only way we could really know whether the treatment is making a difference is to cut you in half and test you. One half we will treat and the other half we won't. Then we can tell whether the treatment worked, and we'll pick the good half and discard the bad half."

She laughed. "Although I don't like the idea of being cut in half," she said, "I wish that were possible."

"I do too," I said.

7

JONAS SALK TRIES TO CURE MS

When I treated Gloria Edwards with plasma exchange, I hoped that the treatment would have dramatic effects. Even though I knew the chances were slim, it was worth a try. Whenever a new treatment is first given, in the back of our minds there is the hope that a true breakthrough will be found, that this time we will somehow capture the brass ring. There are many instances of triumphant breakthroughs in medicine, of brass rings successfully captured, but few are as deeply ingrained in our consciousness as the story of Jonas Salk and the polio vaccine. And associated with the brass ring, of course, is the Nobel Prize. Of all the recognition given in science and medicine, nothing is higher than the Nobel Prize. It is the pinnacle of achievement—a form of sainthood reserved for someone who has gone for the brass ring and grabbed it, someone who understands the world in a way others do not.

More than one patient had asked me directly, "Where is the Jonas Salk for MS?" One patient told me that she thought I would be the Jonas Salk of MS, and many patients have told me that they were confident that I would win a Nobel Prize and that they wanted to join me on my trip to

Stockholm when the time came. Many MS researchers have told me that they have received the same praise. It is embarrassing and uncomfortable to be told this by patients, because I know how complex MS is and that there is unlikely to be a simple answer. But we do know that one day there will be a cure for MS, and it will be discovered by researchers and doctors like myself. Whether we will win a Nobel Prize is another story.

Although Jonas Salk's name is linked to one of the great accomplishments of modern medicine, few people, even scientists, know that Salk never won the Nobel Prize for his work on polio. That honor went to John Enders, Tom Weller, and Fred Robbins, who discovered how to grow poliovirus in tissue culture. Enders was able to grow poliovirus because he applied what he had learned about growing a different virus (chicken pox) to the problem. In the eyes of the Nobel Prize committee, the real breakthrough in curing polio was the ability to grow the virus in the lab, which paved the way for creating the vaccine. Salk merely applied someone else's discovery to the problem. Nonetheless, it was Salk's bold approach and single-mindedness that led to the vaccine trials and the beginning of the eradication of polio, which may soon be accomplished. Was the Nobel Prize committee right or wrong? Everyone in the world knows Salk's name; no one knows the names of Enders, Weller, and Robbins. When it comes to polio, does it matter who won the Nobel Prize?

I have framed a number of quotations and hung them in my office. One of them is from John Enders, in which he describes how discoveries are made in science. "We are dependent on one another," Enders wrote. "It is an erroneous impression that scientific discoveries are made by an inspiration or a thunder clap on high. As a rule the scientist learns from the observations of his or her predecessors and shows individual intelligence (if any) by the ability to discriminate between the important and the negligible and by selecting here and there the significant stepping stones that lead across the difficulty to new understanding." Enders' quote is reminiscent of Newton saying that he could see as far as he could because he stood on the shoulders of giants.

Science's achievement in controlling polio has been both a blessing and a curse for modern science. Harold Varmus, the former scientific director of the National Institutes of Health and himself a Nobel Prize winner for his work on retroviruses, explains, "In a sense we've inherited the mantle of responsibility to do what we did for polio for every other disease." Consider cancer. The "war on cancer" was initiated under President Richard Nixon in 1971, but we still have no cure. The explanation is simple. Polio is a single disease caused by a known agent. Diseases such as cancer and multiple sclerosis are much more complicated and thus much more difficult to understand and treat. But in the back of everyone's mind, both scientist and patient, is the thought that perhaps there *is* a single key, a single switch that is flipped the wrong way in MS patients. If such a switch exists and can be found, MS could be cured. That might even be worthy of a Nobel Prize. However, there may never be a Nobel Prize for MS. Tomas Olsson, a Swedish MS researcher who sits on the Nobel Prize committee, reminded me during one of our recent discussions that only three people can be given the Nobel Prize for any one discovery. With the complexity of biologic processes such as MS and the large number of scientists working in the same area, there have been Nobel Prize quality discoveries, but because many more than three people are involved, no prize could be given.

Jonas Salk, however, did go for the brass ring with his MS experiment. When he tried to develop a vaccine for multiple sclerosis, he was not as optimistic as he was with his polio vaccine, but he took a bold step and carried out an experiment that perhaps only a man with his reputation could undertake, although he was not the first to try the approach.

Salk began his attempt at an MS vaccine in 1978 and 1979, just as I was setting up my first laboratory (it was in a research building named for John Enders, which I took as a good omen). After polio had been eradicated from most of the world, MS inherited the mantle as the leading paralyzer of young people. Indeed, some of those who worked on curing polio later found themselves working on MS. Harry Weaver, who was research director of the National Foundation for Infantile Paralysis and who encouraged Salk to work on polio, was also research director of the National Multiple Sclerosis Society between 1966 and 1977. And Tom Rivers, the virologist credited with describing EAE, had previously headed the research committee for the National Foundation for Infantile Paralysis.

In 1979 the two main prevailing theories regarding the cause of MS, a virus or an abnormal immune response, were being intensively studied. Despite their lack of success, scientists were furiously searching for an MS virus, and at the same time they were testing treatments in the animal form of MS, EAE. The EAE line of research had reached a crucial juncture. Although no one knew what caused MS, there was increasing interest in the hypothesis that it was caused by an abnormal immune response against MBP, the protein used to induce EAE. There was growing pressure to test the hypothesis.

There are analogies to what Salk did in trying to develop an MS vaccine and what he did to develop one for polio. In polio, the organism that causes the disease is the poliovirus itself. The virus enters the body, attacks the spinal cord, and causes paralysis. The poliovirus attacks the spinal cord because there are specific receptors for the virus in the nervous system. The scientific theory behind vaccination against poliovirus is simple: make a weakened virus that cannot cause disease, inject it into people, and the body will then mount an immune response against the virus. This immune response consists both of antibodies and T cells that can kill or neutralize the virus. When real poliovirus enters the body, the immune system recognizes it immediately and neutralizes it, preventing it from attacking the nervous system.

In MS, however, there is no virus. So what can we vaccinate with? How do we make an MS vaccine? What in MS works the same way as the poliovirus? In MS, the theory is that one's own T cells are the culprit; instead of poliovirus, one's own T cells attack the brain and spinal cord and cause paralysis. It is point number one of the twenty-one-point hypothesis.

Thus the theory behind Salk's vaccine against multiple sclerosis was to shut down the T cells that were attacking the brain and spinal cord. The Salk vaccine for MS was based on all that was known at that time about the animal model for MS. In fact, Salk performed an experiment in people similar to the one I performed in guinea pigs in Barry Arnason's lab when I was a neurology resident in 1973. Just as the poliovirus targets receptors on the spinal cord, in MS the postulate was that the body's own T cells target specific proteins in the brain and spinal cord. At the time of Salk's study, the hypothesis was that the brain protein being attacked by T cells was myelin basic protein or MBP, the most-studied brain protein in MS and the one I had purified in Arnason's lab a few years before. We knew that we could suppress EAE in animals by desensitizing or paralyzing T cells targeted to MBP. We did it by injecting MBP in soluble form. Thus the soluble form of MBP was like the weakened form of poliovirus that Salk had used in his vaccine.

However, there was a crucial difference between vaccinating with poliovirus to prevent polio and injecting MS patients with soluble MBP to cure MS. It was known that poliovirus causes polio. There was no evidence that MBP had anything to do with MS; it was only a hypothesis derived from the animal model. At the time of Salk's MS experiment it had not been clearly demonstrated that MBP-specific T cells even existed in the blood of MS patients.

Salk's polio vaccine was built on Enders' discovery of how to grow the virus. Salk's MS vaccine was built on the discovery of EAE by Rivers in 1933, on the isolation and characterization of MBP by a number of scientists beginning with Kies and Alvord in 1959, and by the demonstration that injection of soluble MBP in animals could prevent the animal model of disease. Indeed, once EAE was discovered, it became fairly easy to find treatments for it. As early as 1949 Ferraro and Cazzullo successfully treated EAE by injecting brain proteins in a weak, rather than strong, adjuvant. This was a forerunner of Salk's attempt to treat MS by injecting soluble MBP. Alvord has tracked the question of whether EAE was an appropriate model for MS or whether MS was caused by a virus by making a graph. Like the stock market, that theory has had its ups and downs over the past fifty years. Nonetheless, for Salk the logic to treat MS with soluble MBP was clear: if EAE indeed was the appropriate animal model for MS and MS was caused by T cells reactive to MBP, injecting patients with MBP could cure MS.

Salk's was not the first to attempt to cure multiple sclerosis by injecting MBP. In July 1973 a group led by Berry Campbell reported on a trial in which they injected sixty-four MS patients in a double-blind trial to test the efficacy of desensitizing MS patients to human MBP. The problem, of course, is that injecting MBP itself could theoretically exacerbate MS. Remember, injecting material containing MBP multiple times is how Rivers first demonstrated EAE in monkeys. The central question then became which type of MBP to inject and how much to inject. The MBP used by Campbell was prepared from human brains obtained at autopsy from people who had no history of neurological disease and was injected once a week for thirty months. The results? There were no side effects from injecting the human MBP, but unfortunately no benefit could be demonstrated either, though some patients receiving the MBP reported a feeling of "well-being." Campbell and colleagues concluded that the next step might be in using portions of MBP, or synthetic proteins. Nonetheless, Campbell's group were not encouraged enough by the results to repeat the study.

A second attempt to cure MS by injecting MBP was performed by a group from Belgium headed by Gonsette and reported in 1977. Gonsette's group treated thirty-five patients with the relapsing-remitting form of multiple sclerosis with a preparation of both human and cow MBP. To prevent possible anaphylactic or allergic reactions, patients were given 0.05 mg the first week, 0.5 mg the second week, and, when no complications were observed, they were given a weekly injection of 5 mg over a period of three to eleven months. Most patients were treated for approximately seven and one half months. Like the first plasma exchange trial I carried out with Dave Dawson, the study was not double-blind—both patient and doctor knew the treatment was being given. The results were negative. The treatment did not reduce the relapse rate of MS patients when compared to their relapse rate prior to treatment, and there was no change in lymphocyte reactivity to brain proteins. Gonsette's group concluded that one could treat without major side effects, but there was no therapeutic effect. They ended their article by writing, "Failure of MBP therapy seems worth reporting since when treatment of MS is concerned, even a negative result may be interesting."

It was against this background that Salk began his attempt to cure MS by injecting MBP. Of course, when Jonas Salk embarked on an attempt to cure another paralytic disease, it immediately caught the media's attention. Most science is done outside the public eye. The media focuses on scientific experiments depending on who's doing the experiment and whether a scientific report is published in one of the leading medical journals. This first happened to me when we published the results of cyclophosphamide therapy in MS in the *New England Journal of Medicine* in 1983; we could not avoid a national press conference. Campbell and Gonsette performed their MBP trials and published their results in the medical literature. When Salk initiated his trial of MBP in MS, it was carried by all the news media and reported in the *New York Times* under the headline "Human Tests Seeks Possible Link of Allergy to Multiple Sclerosis." Salk's MS vaccine was approved by the FDA, and Salk turned to Eli Lilly, the Indianapolis pharmaceutical company that had been involved in the creation of the Salk polio vaccine, to manufacture the MS vaccine. The FDA, Eli Lilly, and Salk all cautioned against false hopes for the MS vaccine, and Salk himself stated it would require several years before any conclusions could be drawn. A representative of Eli Lilly emphasized that the trials were to stop progression of the disease and there was no indication that the MBP would repair nerve damage. Another Eli Lilly representative stated the underlying theoretical problem with the vaccine succinctly: "The real problem is to know whether the animal model is true for humans. A lot of things react one way in animals and another way in humans." That remains true to this day.

How did Salk's treatment differ from those of Campbell and Gonsette? The two major differences were that Salk used pig MBP and that he injected it in very large doses. It was a bold step. Eleven patients were treated in an open-label, uncontrolled study that began in January 1978, and twelve patients entered a matched-pair, single-blind, placebo-controlled study in 1979 in which six patients were injected with MBP and six were injected with a lactose placebo. MBP of porcine rather than human origin was chosen by Eli Lilly because of the unlimited supply. In preclinical testing there were no abnormal reactions in guinea pigs or monkeys that received ten times the human dose for three months. They did not develop EAE.

As I noted, the doses used by Salk were large. Campbell and Gonsette injected 5 mg. Salk began with a daily injection of 75 mg of MBP and eventually increased it to 225 mg. Both patients with the relapsing-remitting form of the disease and patients with the progressive form were treated. Salk felt larger doses were needed, as in the animal model smaller doses had only partially suppressed the disease. Perhaps that was why Campbell and Gonsette had failed, he theorized.

A total of nineteen patients were treated. After two weeks of daily injections of MBP, the first patient, a thirty-four-year-old man who had had relapsing MS for three years, had stabilization of his signs, and the next two patients, a fifty-six-year-old man and a thirty-eight-year-old man who had had progressive disease for over ten years, had improvement in symptoms. Based on this, MBP was administered daily for up to two years in these patients and in subsequent patients. However, in fifteen of the nineteen patients there was an allergic skin reaction at the site of the injections that sometimes required the administration of prednisone. In eight patients there was temporary worsening of signs and symptoms of MS with the appearance of the skin reactions. Initially, the worsening raised the question of whether there was something wrong with the batches of MBP being used. For Salk, this brought back memories of his experience with the polio vaccine, when after the successful national vaccine trial, a batch of polio vaccine from Cutter Laboratories was found to be defective and actually caused polio in some vaccinated children. As it turned out, the batches of MBP were fine, so the worsening in these patients appeared related to the MBP itself. One patient with fulminant (severe) MS treated with MBP eventually died. It was not clear whether he died as a result of the MBP treatment or simply because of the disease itself. Salk's experience emphasized the double-edged sword of trying to treat MS with MBP. Although potentially beneficial, it could also be harmful; it depended on how the immune system reacted. In fact, in another study done twenty years later, Roland Martin of the NIH reported a worsening of MS when a fragment of MBP was injected into some patients.

The final results of Salk's MS vaccine were disappointing. Most patients were not helped, and some worsened, although it was not clear whether they worsened because of the vaccine or because of the natural progression of the disease. The treatment did not affect the elevated level of gamma globulin in the spinal fluid of MS patients. That many patients had skin reactions to the injected MBP suggests that they may have been sensitized rather than desensitized by the injections; this pointed to it being their first exposure to MBP, which in turn may have meant that MBP was not the target of the T cells. Salk could have been treating with the wrong brain protein. Point two of the twenty-one points is that there is no single brain protein that is targeted by T cells in all MS patients.

Despite the fanfare when the trial started, the results of the Salk MS vaccine were never formally published in a peer-reviewed scientific journal but rather are contained in books that summarize symposia where the work was presented. In one of these articles Salk writes that while there is one agent that causes polio, there might be multiple agents that cause MS. He restated the difficulty of treating an immune-mediated disease such as multiple sclerosis as compared to an infectious disease such as polio. He also postulated that because they could not reproduce in MS patients the positive effects observed in the EAE animal model, EAE induced by MBP was probably different from MS.

THERE IS A deep-seated human desire for things to be simple, for there to be clear answers. Buy a new car and get the girl. Don't eat carbohydrates and lose weight. This is what is so appealing about the polio story: take a vaccine and no more polio. Unfortunately, neither life nor science is black and white. One of the major challenges in science and in our attempt to cure MS is to make progress in a complex system when the answers may be gray rather than black and white. The gray answers still contain a truth that will lead us closer to the cure, but how do we extract that truth? Thus the road to the cure for MS involves statistics and designing the correct experiment much more than going for the brass ring, even though we are always hoping and planning for the brass-ring experiment. The statistical concept that scientists use to extract biological truths is something called the p value.

"What's the p value?" I asked, as I went over a series of experiments with one of my fellows.

"0.06," he said.

"That's not statistically significant," I said. "We can't publish it."

"But it's so close," he said.

"I know," I answered, "but we still can't publish it."

"What if it was 0.05?" he asked.

"Then we could publish it," I said.

Science is quantitative, and we attempt to measure biological phenomenon precisely. Nonetheless, statistical analysis governs the interpretation of virtually all scientific and clinical investigation, and scientists are governed by p values, which refers to the odds of a biological event happening by chance. A p value of 0.05 means that there is a 1 in 20 probability of something happening by chance. A p value of 0.01 means there is a 1 in 100 probability of something happening by chance.

And a p value of 0.001 means there is a 1 in 1,000 probability of something happening by chance. All of our scientific experiments and all of our clinical studies are considered significant only when the p value is equal to or less than 0.05. A p value of 0.06 has arbitrarily been determined to not be significant, whereas a p value of 0.05 is. A drug trial with a p value of 0.06 as opposed to one of 0.05 can make or break a company that is trying to gain approval for a new drug.

Over the years, there have been stories and jokes that have become part of the lore of the laboratory. They have been told over and over again as each new group of research or clinical fellows join the lab. Those who have been around for many years know the stories well and often simply refer to the punch line when the point of the joke is applicable to a scientific problem or issue we are facing. It is a bit like an inmate in a prison simply calling out the number of a joke and other inmates laughing because they all know the joke book and what joke is on each page.

My favorite story has come to be known as the "funeral joke." It has more than one meaning, and one of its meanings relates to p values and the concept of statistical significance. Another meaning relates to the biology of different stages of MS. Its punch line can also be used in more mundane situations, like when it suddenly becomes clear that adding extra people to the van for the drive to New Hampshire for our winter laboratory ski trip won't work. At our annual holiday party, it has become tradition for myself and Dennis Selkoe to tell jokes. Dennis codirects with me the Center for Neurologic Diseases at the Brigham and Women's Hospital and is a leading researcher in Alzheimer's disease. All jokes must have clearance so they do not offend anyone, which is almost impossible since virtually every joke has the potential to offend someone. But the funeral joke has always gotten clearance. It goes like this.

There was once a family in which a dear uncle had died and the man's son, niece, and nephew went to the funeral parlor to discuss arrangements for the funeral. The nephew explained how important the uncle was to the family and asked about the best possible funeral. The funeral director assured the family that they provided the most lavish funerals anywhere.

"Tell us about your most lavish funeral," the niece said.

The funeral director took out a large brochure. "Our most lavish funeral is also one of the saddest," the director said in hushed tones. "We use our main sanctuary that seats five hundred. The sanctuary is decorated with one hundred large bouquets of flowers. We play sad music on a giant pipe organ from Germany. We use thirty black limousines for the family and the guests. There are twenty Italian women dressed in black who sing funeral hymns and who cry at appropriate times. And, of course, we guarantee that it will be overcast and raining on the day of the funeral."

"How much does that cost?" the son of the deceased asked.

"$50,000," the funeral director said solemnly.

"Our family is not one of great means," the nephew said. "What type of funeral do you have for $25,000?"

The funeral director brought out another brochure. "This is one of our middle-level funerals," he said, "but be assured, it is done tastefully and is still a very sad funeral. We put up a partition and only use part of the main sanctuary. Because of this, we can't use the German pipe organ, but there is a smaller organ from Philadelphia that we use and that sounds fine. The number of flowers is cut in half, but you can't tell because the sanctuary is smaller. There are only eight Italian women dressed in black to sing hymns and cry, and we use seven limousines. Of course, for $25,000, we can make no guarantees about the weather. In fact, the sun will probably be shining."

The son conferred with the niece and nephew. "Do you have any funeral plans for $10,000?" he asked.

"You are beginning to stretch things," the funeral director said with a shrug, "but we can do it, and we guarantee it still will be sad. We must move the funeral to our small sanctuary and there will only be six bouquets of flowers, but we will put them around the casket in a tasteful way. There is no organ in the small sanctuary, but my son will bring in his synthesizer that sounds like an organ. I can only provide three Italian women dressed in black to sing hymns and cry, but I'll give you my three best Italian women. We are down to two limousines, but we will be able to accommodate as many as fit in two large stretch limousines," the funeral director said, wiping his head with a handkerchief.

The son thought for a second, conferred with the niece and nephew yet another time, checked a small notebook he had in his pocket and then asked somewhat sheepishly, "What can you do for $5,000?"

The funeral director looked down at his brochures, then up at the son, thought for a second, and said, "Sir, at $5,000 it starts to get funny!"

What is the point of the joke? Why has it become so popular in our laboratory? The funeral joke describes a basic scientific concept. Something may change quantitatively, in a step-by-step fashion, but at some point the quantitative change actually becomes qualitatively different. It is no longer just a less lavish funeral; at $5,000, the basic property of the funeral—it's being a sad affair—changes. At $5,000 it becomes funny.

MS patients who have relapsing-remitting disease experience different frequencies and severity of attacks. Some patients may have one attack and never have another, others may have attacks only on occasion, and still others may have them frequently. When is a line crossed? When does the number of attacks signify something qualitatively different about the disease process as in the funeral joke?

Over time we have learned that the number of attacks in the first few years of the illness has prognostic value about the future course of the disease. There is a different prognosis for someone who has two attacks in the first years of illness versus someone who has seven attacks in the first years of illness; the latter patient is likely to do more poorly. There are people who have one attack and never have another, and in this case, if there is no disease activity on MRI scan, a diagnosis of MS cannot be made. At our MS center, we call this "singular sclerosis" or an isolated demyelinating episode. It doesn't officially become MS until there is a second attack or new disease activity on MRI.

The same concept applies to the relapsing-remitting and progressive forms of the disease. Something qualitatively different happens when the disease changes from the relapsing-remitting form to the progressive form. Disability in MS is most frequently associated with the progressive form of the disease. In progressive MS, the clinical disease course changes its nature, the prognosis is poorer and treatment is less effective.

The funeral joke also applies to p values and statistical significance. It's another example of the quantitative becoming qualitative, of the sad funeral becoming funny. Something "funny" or significant occurs when the p value changes from 0.06 to 0.05. Only, in this instance, the qualitative change has been artificially assigned. We arbitrarily decide that the cutoff point is when it is at 0.05. Actually, in order to do statistics and get a p value, you have to compare one thing to another. That is the basis for clinical trials and the approval of new drugs for MS, and that is the basis for discoveries in the laboratory. For our study of the chemotherapy drug cyclophosphamide in MS that we will discuss in the next chapter, we compared the number of patients that were stable or improved in the cyclophosphamide plus ACTH group (16 out of 20) to those that were stable or improved in the ACTH group alone (4 out of 20). The p value was 0.002; that is, the likelihood that we would get such a result by chance was 1 out of 500. Thus we were able to conclude that cyclophosphamide had a positive effect in the patients we studied.

A crucial part of making discoveries is knowing how to ask the right question, a question that will uncover differences and result in a positive experiment and a statistically significant p value. The bigger the difference between the two groups, the smaller number of times one has to do the experiment, and the smaller numbers one needs for the experiment to be positive, to have a significant p value. This concept relates to the design of experiments and asking the right question when one does an experiment in the lab or when one tries to prove that a new drug is working in MS. One has to ask the right question.

Imagine you want to distinguish between two runners, myself and Bill Rodgers, who I watched win a number of Boston Marathons when I was in my early days of learning about MS. Now Bill Rodgers can run the twenty-six miles from Hopkinton to downtown Boston in two hours and thirty minutes. I'll tell you about myself. As much as I would have liked to, I have never run a marathon. I jog two to three miles twice a week and I run six- to seven-minute miles. I am in good shape, I swim and play squash, but the longest I've run is ten miles, which took me an hour and a half.

Let's assume you have never seen me or Bill Rodgers before and want to do an experiment that identifies who is the Boston Marathon champion. I wear a T-shirt with the letter A and Bill Rodgers wears a T-shirt with the letter B. You design the experiment and then someone brings you a sheet of paper with the results. Let's assume your first experiment is to ask whether each runner can run five miles in one hour or less, and you had the runners attempt it every day for ten days. At the end of the experiment, you received a printout that reads: runner A=10/10 and runner B=10/10; both did it successfully on ten consecutive days. So you design a second experiment in which you ask that each runner try to run fifty miles in three hours or less, again on ten consecutive days. This time the printout reads: runner A=0/10 and runner B=0/10. Again, no difference. Neither runner can do it. You performed two experiments that asked the wrong question and thus did not uncover the underlying biological truth. Now, if you designed an experiment in which you asked whether the runners could run thirteen miles (half the length of a marathon) in seventy-five minutes or less, the printout would read A=0/10 and B=10/10. Eureka! The marathon runner is B. Furthermore, if you repeated that experiment several times, you'd get the same answer every time. Interestingly, there are experiments you could perform that would give intermediate answers. For example, how many times in a row can runner A or runner B run one mile in six minutes? Printout: "Runner A=6/10 and runner B=10/10 times." There is also an experiment you could design that changes one variable until the correct answer is obtained. For example, ask whether each runner could run thirteen miles over a range of times—seventy-five minutes, eighty-five minutes, ninety-five minutes. There would be times that clearly distinguished runner A from B, times that partially distinguished runner A from B, and times in which there was no difference. Additionally, one would obtain different p values for each experiment. The p value for the second experiment (0/10 vs. 10/10) would be highly significant, p=0.0001 (it would only happen by chance one in ten thousand times), whereas the p value for the third experiment (6/10 vs. 10/10) would not reach statistical significance (the actual p value is 0.1, a one in ten probability of it happening by chance).

In studying a disease such as multiple sclerosis and trying to understand it, we are always asking questions, testing, carrying out experiments. But the questions have to be posed in the correct way. If a possible treatment is not tested on the right group of patients, we may not see positive effects. A drug may act differently in relapsing-remitting patients than in primary progressive patients. It may act differently depending on how old the patient is or how long the person has had the disease. In our study of T cells, if there is a difference in certain T cells between MS patients and people without MS, those differences may only be seen if the cells are cultured for the appropriate length of time. Furthermore, we could be testing the wrong function of the T cell, a function that is the same between an MS patient and a normal individual (in the running example, the ability to run five miles in ninety minutes), whereas a different function (running the Boston Marathon) would clearly show the difference between a patient who has MS and someone who doesn't.

Remember the experiment where my fellow came up with a result that had a p value of 0.06?

"What do I do?" he asked, staring wistfully at the data that led to a p value of 0.06. "I know the effect is real."

"Use more animals per group," I said. "That will increase your chances of getting a lower p value."

"I tried that," he said. "It doesn't work. But I've thought of other things I can do. I can change the strain of the animals. I can give the protein at different concentrations. I can use a different adjuvant. I can culture the cells longer."

As I thought about it, it became clear that with all the additional manipulations, my fellow was no longer looking for a robust biological truth but was manipulating the system just in order to get a positive result, even though it would be of questionable biological significance. He had crossed the line.

"There are too many things you are trying to change," I said. "I'd stop. It's starting to get funny."

In science one has to know when to stop doing an experiment, when one is pushing the system for an answer even though there is no useful truth to be discovered. Like it says in the Kenny Rogers song "The Gambler," you have to know when to hold 'em, know when to fold 'em. P values help out, but there are other considerations. A failed experiment does not mean the underlying theory is wrong. It simply may mean the problem was approached in the wrong way or it may have been too early to get the answer because all the tools were not in place. Remember, trying to reach the moon before there were computers was impossible.

Therefore, just because Salk and others did not succeed in the late

1970s with their experiments on desensitization to MBP did not mean the approach was invalid or that immune modulation using MBP might not work one day. I discussed these issues with Salk at scientific meetings and last spoke with him on October 16, 1994, in New York at a fundraiser for the Weizmann Institute of Science. I thought of the pictures of Salk as a young man in Pittsburgh, with black hair, growing poliovirus in hundreds of large roller bottles. In 1994 he had small amounts of wavy white hair combed straight back and a lined face, but was thin and animated. In recent years he had begun work on trying to develop an AIDS vaccine. He had been asked about polio so many times, few people discussed MS with him.

"It's been over ten years since your experiments in trying to cure MS by injecting patients with MBP," I said. "Why do you think they didn't work?"

"First of all, I'd be careful in using the word *cure,"* he said. "When we tested the polio vaccine, I thought we had a chance to eradicate polio. It was all right there. We never thought we were going to cure MS in our pilot trials. We were testing the hypothesis that MS was analogous to EAE, and I wanted to make sure that we injected enough MBP to see an effect. If our results injecting MBP into MS patients were dramatically positive, it could have led to larger trials and then perhaps to a cure. But MS is so much more complicated than polio. Polio is a single disease caused by a single agent. I think MS is a syndrome with multiple causes."

We now know that there are a number of brain proteins besides MBP that can cause EAE in animals, and any of the proteins could be the target of an autoimmune attack in MS. That, in fact, is point two of the twenty-one-point hypothesis.

"When you began injecting MS patients with MBP," I said, "two patients began to improve, and later two other patients appeared to have good responses. As I remember, one of the patients in your trial had been confined to a wheelchair, unable to support his own weight, and after several weeks of daily injections of MBP he could walk with the aid of parallel bars. What did you think when you saw that?" I asked, as I thought back to the response of the first person we had treated with plasma exchange.

"That was a man in his early thirties with the progressive form of the disease," Salk said. "We were excited, but it was only an isolated observation. It immediately became clear that we hadn't found a magic cure. In fact, some of the patients in placebo group stabilized after a year of treatment," he said, "though none of them improved."

"Why didn't you do more trials?" I asked.

"We were afraid of adverse reactions from injecting too much MBP, and although there was a suggestion that some patients might have improved, the response was not strong enough."

Although this was true, the scientific community hadn't given up on the idea of desensitizing or tolerizing myelin antigens. Our group was in the midst of trials in which we were feeding myelin to MS patients to desensitize them via the gut, and a group from the Weizmann Institute was testing a synthetic polymer that they believed mimicked MBP. Ironically, we were at a Weizmann fund-raiser, and the first drug to be approved for the treatment of MS that was based on the theory of MBP desensitization began from work done at the Weizmann Institute in 1971.

We were sitting at a table with Torsten Weisel, who had won the Nobel Prize with David Hubel for discovering basic mechanisms by which the visual system worked.

"Can I ask you a personal question?" I said to Salk.

He looked at me for a second. "Sure, go ahead."

"How do you feel about not winning the Nobel Prize for your work on polio?" I asked, out of earshot of Weisel.

Salk didn't seem bothered by the question and answered immediately. "In the end it makes no difference," he said. "I'm thankful to have done what I've done."

"How do you think we will cure MS?" I asked.

"Step by step," he said. "It is a complicated disease."

What can we conclude from Salk's experiments? Salk took one of the basic premises regarding the autoimmune theory of MS—that MS is caused by T cells that attack a specific protein in the brain, called MBP—and tested it in humans. His tests failed because of an incomplete understanding of the disease and of basic concepts in immunology. Furthermore, at the time of his experiments there was no MRI imaging to allow us to see what was actually happening in the brain or sophisticated immunological tools to test responses to MBP. Furthermore, many of the patients he treated may have been too far along in the disease process to be helped. Unlike polio, where one can measure an antibody response against the vaccine, it was not known what immune measure to test in multiple sclerosis. It was a good try, but it was probably too early in our understanding of the disease. As I might have told my childhood friend Normie, Salk was trying to fix the car without understanding everything under the hood. Nonetheless, there were other ways to test the immune theory of MS, and soon after I established my laboratories in the Enders research building, we embarked on a major clinical trial to test whether we could affect MS by suppressing the immune system.

8

CHEMOTHERAPY FOR MS

IN 1960 H. HOUSTON MERRITT, head of the Neurological Institute in New York and one of the discoverers of Dilantin, a drug that revolutionized the treatment of epilepsy, wrote this: "If you want to ruin your scientific career, publish something positive on the treatment of multiple sclerosis."

On January 23, 1983, I did just that, and I did it with the whole world looking on. Sitting in front of television cameras and in front of photographers whose shutters fired whenever I turned my head, I announced the results of a clinical trial that was to be published the next day in the *New England Journal of Medicine.* We reported that cyclophosphamide, a commonly used chemotherapy drug, plus ACTH, a hormone that stimulates the adrenal gland to release cortisone, could stop progressive MS in 80 percent of patients for at least a year.

I had no choice but to sit in front of the press. Arnold Relman, editor in chief of the *New England Journal of Medicine,* had phoned a few weeks before to tell me, "Dr. Weiner, I think you should know that your article on multiple sclerosis is bound to generate press interest." He reminded me that although the journal officially comes out on Thursdays, they send out copies of that week's issue to the press on Monday so they can prepare their stories, though the press is forbidden to release those stories until 6:30 p.m. on Wednesday (in time for the national news). Nowadays, it is common for physicians to first learn of a claimed medical breakthrough from patients who heard it on the news or read it in the papers.

I prepared a statement that I read to the press at the start of the news conference; I urged caution. I stated that the treatment was not ready for routine clinical use. I emphasized that we had not found a cure, but perhaps had made a first step toward finding an effective treatment. I tried to be as negative as possible, because I didn't want to create false hope. But underneath, I was ecstatic about our findings and felt we had truly made a breakthrough. A breakthrough in a disease that was one of the most difficult to study.

STEVE HAUSER BECAME a fellow in my laboratory after finishing his neurology residency at Massachusetts General Hospital in 1980. He came over to meet me while I was still in Bernie Fields' lab, and we began to collaborate on immunological studies of MS in 1978, while he was still a resident. I remember Steve, still in his tie and white coat from hospital rounds at Mass General, preparing blood in a tiny lab that belonged to Ken Ault just adjacent to Bernie Fields' lab. We were studying natural killer cells in MS and sending the samples to Ellis Reinherz at the Dana Farber Cancer Institute, where they were checking for suppressor T cells. The samples we sent resulted in a publication in 1980, but it was our collaboration on the use of cyclophosphamide in MS, published three years later in the *New England Journal of Medicine,* that created the biggest stir.

Like myself, Steve had a passion for wanting to understand and treat multiple sclerosis and has spent his career studying MS—most recently developing a model of MS in the marmoset and doing studies on the genetics of MS as chairman of neurology at the University of California, San Francisco. His interest in MS began when he treated a young woman, a beauty queen and law school graduate, whose life was forever altered by MS. He spoke to Raymond Adams about MS, just as I had when I was a resident and attended my first neuropathology conference in 1971 at Mass General. Steve distinctly remembers Adams telling him, "MS can be broken. You can figure it out." Because Barry Arnason had left Mass General for Chicago, there was no one at Mass General currently doing MS research, and he heard about a young researcher in Bernie Fields' lab interested in MS. Like Adriano Fontana, my very first fellow, Steve interviewed with me in my little office in Bernie's lab that had mouse cages on the shelves. It was the year before I moved to the Enders research building. I kept the cages in my office because I wanted to follow my experiments very closely, something that impressed Adriano, and once I even took my mice home to give them a treatment in the middle of the night. Today, it would be impossible to keep cages of mice on one's desk, one has to enter mouse facilities dressed in sterile gowns. Steve Hauser asked Bernie Fields whether it was advisable to work in the lab of someone as junior as me. He was given the okay and ultimately we did more than okay together, publishing twenty-one peer-reviewed articles and a number of reviews.

In 1980, when Steve joined the lab, MS therapeutics had no clear direction. Most physicians treated MS with prednisone or ACTH. Salk was in the middle of his experiments injecting MBP. Treatments that would be approved by the FDA for the treatment of MS in the 1990s—interferons and an MBP analog called Cop-1—were still in their infancy in clinical testing. One of the reasons there was no clear direction for MS therapeutics is that there was still a major debate on what caused the disease, and of course treatment would differ depending on the cause. Some felt it was a virus and it was just a matter of time before the virus would be found. Others felt the key to the disease was the immune system, but there was debate over what the flaw of the immune system was.

If MS was like EAE, then the immune system was overactive and had to be suppressed. Indeed, there were a large number of investigators testing the effect of immunosuppressant drugs on MS. However, others believed that there was a defect in the immune system and that for whatever reason it couldn't clear viral infections from the body. This led to trials of a compound called transfer factor, which was extracted from the lymphocytes of normal individuals and transferred to MS patients, where it would beef up a weakened immune system so that it could handle viral infections. In 1976 there was a case report of a twenty-three-year-old woman with MS in South Africa who was given six injections of transfer factor at two-week intervals, to improve immunity to mumps virus. The authors reported that the patient's symptoms improved dramatically and she had no exacerbations (relapses) for ten months, but cautioned that "MS is a disease of exacerbations and spontaneous remissions, and despite our promising findings we wish to refrain from hailing mumps transfer factor as a possible cure of this notorious disease." However, in the 1980s large controlled studies showed no effect of transfer factor on multiple sclerosis.

Trials of interferons, the first drugs to be approved by the FDA for the treatment of MS, were also begun in the early 1980s. Those trials were based on the belief that there was a defect in the ability of MS patients to handle viruses (although interferons had many actions and, depending on the interferon, could affect the immune system by either boosting or suppressing it). My first independent grant to study MS came from the Kroc Foundation and was for the study of natural killer cells. Steve and I were one of the groups that found a defect in natural killer cells in MS patients, which was part of the basis for testing interferon.

We had just completed our study of plasma exchange and were excited about its potential to help MS patients. Steve was finishing his residency at Mass General and spending elective time in the lab. Being clinically based, Steve took care of MS patients and, like myself, was confronted with questions of therapy and what was ultimately best for the MS patient: to suppress the immune system, to stimulate it, or to help it fight off viral infections? As often happens in medicine, our interest is frequently sparked by encounters with individual patients, and we learn from them. Steve's interest was sparked by taking care of an MS patient who developed breast cancer. Her MS was quite active when the breast cancer was diagnosed; she had frequent relapses and was rapidly developing disability. Because of the breast cancer, she was treated with chemotherapy, and with the treatment, a remarkable thing happened: her MS got better. She stopped having relapses, and her clinical condition improved.

Steve then did an extensive review of the literature and of all the types of therapy that were being tested in MS. One of the major approaches to finding a treatment for the disease was to use drugs to suppress what was thought to be an overactive immune system. Steve found that of all the immunosuppressive drugs currently used, the one with the strongest effects was cyclophosphamide, which was used as a chemotherapy drug to treat a large number of cancers, although it had also been used in inflammatory diseases that were not cancerous. This was what had been given to his patient with breast cancer. Thus, just as it became obvious to investigators to use steroids in MS if it was an inflammatory disease, it also became clear to several investigators that drugs that suppressed the immune system might also help. Two investigators had studied cyclophosphamide in Europe, Otto Hommes and Richard Gonsette. (Gonsette was one of the investigators who had also worked with MBP prior to Salk's attempt.) Although their studies were uncontrolled, they reported often dramatic effects in some of the patients they had treated.

We sat in one of the conference rooms in the Enders research building and talked about the possibility of doing a clinical trial. Dave Dawson joined us, as did people from the blood bank at Children's Hospital. With Jim Lehrich at Mass General, Steve had already treated two MS patients with cyclophosphamide, using a protocol modified from that of Hommes and Gonsette. Lehrich said they had seen positive results in the two patients treated, though he was worried about side effects from such a strong drug. We ourselves were debating what to do next in our studies of plasma exchange.

Steve and I had spoken before the conference and we decided to present the idea of doing a randomized trial that involved both cyclophosphamide and plasma exchange. "The problem we face," I said, "is that there is no treatment for MS apart from ACTH, and we are confronted by watching our patients worsen despite ACTH treatment."

"The work of Hommes and Gonsette suggests that immunosuppression with cyclophosphamide works," Steve said, "and Howard and Dave have results that plasma exchange may help."

Thus the idea was born: take patients who were suffering the most and hit them hard for a short period of time by suppressing the immune system or giving them plasma exchange. Suppressing the immune system in MS wasn't an original idea, but we felt we had something new to test with the plasma exchange, and for the first time we would attempt a controlled trial where one treatment was compared against the other.

"What type of patients should we treat?" Dave Dawson asked. After a prolonged discussion, we decided to choose patients who were in the chronic progressive phase of their illness. We reasoned that those whose disease was inactive or who had the relapsing-remitting form of the disease would be more difficult to assess and that the chronic progressive form was more malignant and predictable. Most patients with the progressive form of MS require a cane within two years of the onset of the progressive phase, and deficits in patients with progressive MS rarely improve on their own. For the study, we chose patients with marked worsening in the nine months before entry, as defined by the Kurtzke disability scale; no steroid therapy within a month of entry or other immunosuppressive therapy within a year; and the patient had to be able to walk in the two months before admission.

"What will the control group be?" Jim Lehrich asked. We knew what two of the treatment groups would be. One group would receive cyclophosphamide the way Steve gave it at Mass General, and another group would receive plasma exchange.

"How about no treatment, or giving an infusion of saline?" I said.

"I don't see how you can do that," Dawson said. "Imagine taking people who are rapidly worsening with MS and telling them they may get nothing. Because of the treatments we are giving, the patients will know what they are getting. They will be hooked up to a machine or getting infusions of cyclophosphamide. If people know they are getting no treatment, they will expect to do worse."

"A control treatment could be ACTH," I said.

"Many of them have already had ACTH and it didn't work," Steve said.

"I have an idea," Dawson said. "There is a new form of ACTH that is given by infusion rather than by injection. It may be more potent than the intramuscular ACTH that people received in the past. The control group would be put in the hospital and get an IV just like the cyclophosphamide groups, and they'll be treated with something they hadn't been treated with before."

"I like it," I said. "Each patient will be getting something that might help."

"I think we should give every group the IV ACTH," Jim said, "to maximize the treatment responses and to make sure everyone receives a form of conventional therapy. What if the new IV ACTH is as good as cyclophosphamide? We will also then be able to determine the degree to which the other treatments are better than ACTH."

"It's also theoretically possible that ACTH could make people worse," Dawson said, "so it's good to have it in all the groups. If the

ACTH group gets worse and the other two groups don't, then it's not likely to be the ACTH."

Thus, all patients received IV synthetic ACTH, which was a stronger form of ACTH than that given by injection into the muscle. The first group received nothing else. The second group received five plasma exchanges over a two-week period and oral cyclophosphamide, because other studies had suggested that immunosuppression was required for plasma exchange to have a beneficial effect. The third group received intravenous cyclophosphamide in the hospital for ten to fourteen days in order to lower their white blood counts from normal levels of 5,000 to 8,000 to between 1,000 and 2,000 (the white blood counts would return to normal approximately a week later). Cyclophosphamide's major toxicity was on the bladder, so we gave large doses of IV fluids to flush the drug out of the bladder. If someone had urinary problems, we inserted a urinary catheter.

We discussed the possibility of performing a blind trial, but we felt it was impossible. The patients would obviously know which treatment was being given. One group would be hooked up to a plasma exchange machine, and the IV cyclophosphamide group would temporarily lose all their hair. We also didn't feel we could blind the physicians either. Nonetheless, because patients who were eligible for the study were randomly assigned to one of the three treatments, it would be one of the first controlled studies of immunosuppression in MS. And yet, the fact that the study was not double-blind nor placebo-controlled would haunt me for the next twenty years.

Patients were hospitalized for two to three weeks either at the Brigham and Women's Hospital or at Massachusetts General Hospital. As before, plasma exchange was performed at the Children's Hospital. We used the disability scale designed by Kurtzke to determine whether the patients had worsened.

"I don't like the Kurtzke scale," I said to Steve as we put together our clinical forms to follow the patients. The Kurtzke disability status scale (or DSS, as it was called then) assigned precise values to different parts of the neurological exam (eye movement, coordination, sensation) and assigned a score based on a composite number. Later the scale was expanded and called the expanded disability status scale, or EDSS. People with an EDSS of 0–3 had mild disease; an EDSS of 6 meant that a cane was required for walking; and those with an EDSS of 7 needed a wheelchair. People with an EDSS of 4 didn't stay at that level for long—it was a transition phase.

For most of our patients in the progressive phase, the major problem is walking. The EDSS did not capture worsening in ambulation very well, one of the major causes of disability in MS. Thus, we created an ambulation index, in which we simply measured the time it took a patient to walk twenty-five feet and whether they walked it unaided, used a cane, or needed crutches. We created it by walking down the hall ourselves and field-tested it on a series of patients prior to the study. With the publication of our study, the ambulation index became a standard measure of MS disability. In our new MS center, which opened in 2000 and combined patients from Brigham and Women's Hospital and Massachusetts General Hospital, there is even a twenty-five-foot inlaid red carpet in the hallway, where the ambulation index can easily be measured. At our clinical meetings over the years Dave Dawson has said that he doesn't like the EDSS, and several other MS centers have created their own neurological rating scales for MS. Nonetheless, the EDSS has remained a basic measure of MS disability until this day and is usually part of every major clinical trial. MS physicians know that when the EDSS rises above 3 it is generally a bad prognostic sign for the patient.

There were three time points that we used to assess response to therapy: prior to treatment, at six months following treatment, and at twelve months following treatment. In addition to the actual scores on the EDSS and ambulation index, we classified the patients as stable, improved, or worse, depending on changes in these scores. A treatment failure was someone who had worsened one or more points on the disability scale to the level of 6, where a cane was required to walk. All patients had a spinal tap before the study and investigation of T cell subsets before and after treatment. MRI imaging of the MS process in the brain did not exist at the time of our trial, so all we could go by was the clinical response.

For those patients receiving cyclophosphamide, we prepared white index cards that recorded the daily white blood cell count and the amount of cyclophosphamide they received. We were worried that we might cause too much immunosuppression with cyclophosphamide, resulting in infection because of low white blood cell counts, but on the other hand, if we didn't give strong enough treatment, we might not see a positive effect. Remember, these were people with severe progressive disease, and we were trying to replicate the results of Hommes and Gonsette, who specifically tried to lower the white blood count. "Be careful you don't kill someone," Lehrich had said to Hauser at the beginning of the trial. In order not to give too much cyclophosphamide too fast and to control the amount given, the daily dose was divided into four parts and given by IV infusion on the hospital research floor.

There was also a theoretical immunological risk in giving cyclophosphamide. Cyclophosphamide was one of the first drugs that had been shown to suppress EAE, the animal model of MS, and it did so dramatically. Nonetheless, in another animal model of an autoimmune disease, the mouse model of diabetes, the disease could actually be induced by injecting cyclophosphamide in the right strain of mice. Paradoxical responses after therapy can happen, and there have been trials of drugs shown to help in animal models that made MS worse in people. There were also reports that suppressor cells were selectively destroyed in animals given cyclophosphamide, and in accordance with the autoimmune theory of MS, there were reports of a defect in suppressor cells in the disease. Nonetheless, despite these theoretical concerns, there was no evidence from the studies of Hommes and Gonsette that cyclophosphamide made MS worse in humans, and Steve didn't observe it in some of the patients he treated in his pilot studies at Mass General. So we concluded that the risk was more theoretical than actual.

The study went surprisingly well. There were no major or unexpected complications of therapy, and there was no evidence that cyclophosphamide made patients worse—quite the contrary, in fact.

"I think we are beginning to see something," Steve said one day.

"What do you mean?" I asked.

"In the cyclophosphamide-treated patients," he said, "when their white blood counts reach their lowest point, some of the patients say they are feeling better."

"What about those getting ACTH or plasma exchange?" I asked.

"Hard to say," he said.

"We'll only know for sure when we see them at the six-month follow-up," I said.

When we began to see patients for their six-month follow-up, it was clear that something dramatic was happening. Although we carried out formal ratings on the EDSS and the ambulation index, we also spoke about the "telephone index." Those who had received the cyclophosphamide and some who had received the plasma exchange regimen didn't call; they just appeared at their six-month follow-up appointment. By contrast, many of the patients who received only the new form of IV ACTH called before the follow-up because they were worsening. At the six-month follow-up only two of the twenty cyclophosphamide-treated patients had worsened, and eight had actually improved. At six months in the ACTH-treated group, twelve of the twenty patients had worsened, and only one had improved. An intermediate effect was observed in the eighteen patients who had received plasma exchange and low-dose oral cyclophosphamide. I still remember one of the patients treated with IV cyclophosphamide who was a twenty-five-year-old woman whose MS had begun at eighteen years of age. She had been in the progressive stage for a year prior to coming into the study and receiving treatment. When she entered the study she needed a cane to walk twenty-five feet. At the six-month evaluation, she walked twenty-five feet without a cane in fourteen seconds.

We were ecstatic about the results. The study was designed to include seventy-five patients but was terminated after fifty-eight patients were treated when statistical analysis demonstrated that significant differences existed between the treatment groups. We wrote up the results and sent the article off to the *New England Journal of Medicine*. We elected to include the clinical scores for each of the patients in the study so that physicians could see for themselves what had happened to each patient rather than have only composite scores. At one year 80 percent of those who received IV ACTH alone continued to worsen, while in the ACTH plus IV cyclophosphamide group only 20 percent of the patients got worse and eight patients actually improved. In the plasma exchange group, 50 percent continued to worsen. After revisions and a delay so that one-year data was available on all patients in the study, the manuscript was accepted for publication.

Ours was not the only paper on multiple sclerosis published in the *New England Journal of Medicine* on January 23, 1983. There was a paper by Boguslav Fischer, entitled "Hyperbaric Oxygen Treatment of Multiple Sclerosis: A Randomized Placebo Controlled Double Blind Study." The study reported objective improvement in twelve of seventeen MS patients treated with hyperbaric oxygen in a randomized, placebo-controlled, double-blind study, as compared to only one of twenty treated with placebo.

What led to a trial of hyperbaric oxygen in MS? A 1958 Laten and McKay study had suggested breathing a mixture of 5 percent carbon dioxide and 90 percent oxygen might help acute attacks of MS. In 1970 Boschetty and Cernoch had reported small, transient improvement in sixteen of twenty-six patients treated with hyperbaric oxygen; in 1978 Baixe reported improvement in eleven patients; and in 1979 Neubauer reported minimal to dramatic improvement in 91 percent of 250 patients treated with hyperbaric oxygen. These results were not controlled and so were viewed as anecdotal. In Fischer's study, in addition to the improvement following oxygen therapy, at one year of follow-up, deterioration was noted in eleven patients (55 percent) in the placebo group and in only two patients (12 percent) in the oxygen group. At one year there was improvement in the treated group on the disability scale as well. The improvement reported by Fisher's group was mild and appeared to be greater in patients with milder forms of the disease.

How hyperbaric oxygen treatment helped MS was unclear. Neubauer hypothesized that MS was a vascular disease related to decreased oxygen perfusion in the brain. There was also a study suggesting that oxygen therapy had immunosuppressive effects that protected animals from paralysis in the EAE model, though the animals relapsed when the oxygen was discontinued. No patients in the study worsened and the only side effects were minor ear problems and reversible myopia. The authors wrote that their results of a positive though transient effect of hyperbaric oxygen in patients with advanced MS were preliminary and awaited confirmation.

THUS THE *New England Journal of Medicine,* one of the premier medical journals in the world, published two articles on clinical trials in multiple sclerosis in the same issue. As Dr. Relman had warned me, it created plenty of media interest. I was advised to notify the public relations department at our hospital, where I was immediately told that the only way to handle the situation was by a press conference, where I could answer reporters' questions in a controlled environment. It made sense, but it was a bit frightening. It was not something we physicians or scientists were trained to do, and it was a bit embarrassing because being on television was looked down upon by our professional colleagues. It was not felt to be an appropriate forum to discuss science. Nonetheless, I faced the cameras.

I opened the press conference with the following statement: "We at the Brigham and Women's Hospital and our collaborators at the Massachusetts General Hospital and Children's Hospital would like to urge caution regarding our work published in the *New England Journal of Medicine.* Although we have shown stabilization of severe progressive disease for a year, the treatment is not ready for routine clinical use. Other physicians must confirm it, and further tests are needed to assess usefulness and safety. The benefit we have observed lasted only one to three years. We have not found a cure. We view this as a first step in finding an effective treatment for the disease. Although we are encouraged by our results, the treatment is only recommended for experimental programs or patients with severe progressive disease who have not been helped by standard therapy."

Jim Lehrich spoke next and explained the background of the use of immunosuppression for multiple sclerosis and how many trials had been done, although no controlled studies. Steve Hauser was uncomfortable with the press conference as a vehicle to discuss the results and stood on the sidelines.

Then I made another statement that medical science is not done by one person—that it is impossible to do work without collaborating, without depending on what other people find. "Society won't let us publish simply as a group of scientists studying MS," I said. "We are dependent on what everyone else has done before us, and we can't do it alone." Nonetheless, I could not change how the world works, and my picture appeared in the *New York Times* and *Newsweek.* Could I have escaped it? Probably not. And in the end, the public had a right to know about the work once it was published, and someone had to explain it.

Two editorials were published in the *New England Journal* in association with our article and Fischer's. One was entitled "Multiple Sclerosis and the Ingelfinger Rule" and was written by Marcia Angell, M.D., one of the editors of the journal. The second was entitled "Treatment of Multiple Sclerosis" and was written by Dale E. McFarlin, M.D., head of the neuroimmunology branch at the National Institutes of Health in Bethesda, Maryland.

The Ingelfinger rule was established in 1969 by Franz Ingelfinger, who was then editor in chief of the *New England Journal of Medicine.* The rule stated that the journal would not publish a manuscript if it had been published or reported elsewhere (presentations at scientific meetings were exempt). The purpose of the rule was to ensure that articles underwent appropriate peer review and to protect the originality of work published. It also protected the freshness and interest the public had in the work, which helped the journal. When does a scientific finding come into being? Although there are exceptions, it is when the paper is published in a peer-reviewed journal. It is like a baby being born; at the moment of publication it enters the world, even though it may have been incubating for months (and even though it may be wrong). To add to the drama surrounding the birth, scientific journals impose news embargos on articles, and at the same time send the articles to the press in advance so the work can be reported on the official publication date.

According to the editorial, the *New England Journal of Medicine* had received pressure from a number of sources to report the story of hyperbaric oxygen as quickly as possible and release details of the study to the news media. There were suggestions that the journal was trying to suppress the news of a "cure" for multiple sclerosis. The editorial explained that the hyperbaric oxygen manuscript had been reviewed by no fewer than five experts on multiple sclerosis and twice by their statistical consultant, and had undergone two revisions before publication. The editorial commented that if physicians are not able to validate a study at about the same time as their patients hear of it through the media, then to whom can patients turn to put the study in perspective? The editorial went on to speak of the two MS studies in that issue as "modest," careful trials suggestive of a treatment for MS. They did not represent a cure, but pointed to directions into future basic and clinical research. "We are sure our physician readers will transmit this caution to their patients," the editorial concluded. "We can only hope that the news media will too."

However, reporting the news has become very competitive. Who is the first to declare the winner in the presidential elections? Who can provide live coverage of a snowstorm clogging up the roads? The scientific journals compete with each other as well for top rankings, and they allow the electronic media to report findings on the day prior to the printed version, thus creating interest around the work.

Another factor that we had to be aware of in our presentation of the study's findings was competition between hospitals. A number of popular magazines print lists of the best doctors and best hospitals, and hospitals are aware of their rank on these lists. There was competition between the public relations departments of the Brigham and Women's Hospital, the Children's Hospital, and Mass General over where the press conference would be held. Each hospital wanted to host it—the cyclophosphamide originated at Mass General, the plasma exchange was done at Children's Hospital, but most patients were hospitalized at Brigham and Women's. In the end it was held at Brigham and Women's, with a doctor from each hospital answering questions. The Children's Hospital did get their own chance, however. One night, the local news was broadcast from my laboratories. I remember walking across the connector between the Brigham and the Children's and being handed over by one group of public relations people to another group, as if I was crossing state lines and being handed over from one police escort to another. The television cameras wanted pictures, and as I had done in Colorado as a fellow when the Carp virus story broke, we set up colored solutions in test tubes and moved microscopes around. A scientist at a microscope is like a doctor with a stethoscope or a politician with a baby.

I'll never forget one of the questions that was asked of me that night. "Dr. Weiner," the interviewer with a made-up face, coiffured hair, and clear blue eyes asked, "Would you rather have cancer or multiple sclerosis?" Of all the questions I had thought I might be asked, I wasn't ready for that one—it was a sensational question designed to catch the viewers' attention and put me on the spot. In some ways it wasn't completely ridiculous, though, because some patients have said to me that MS is worse than cancer and we were reporting on the use of an anti-cancer drug to treat MS. We do ruminate about whether one illness is better than another: would you rather be blind or deaf, unable to walk or unable to use a hand, live to eighty with a chronic illness or die healthy in a car accident at fifty? Of course, we know it is better to be healthy, young, and rich rather than sick, old, and poor. Would I rather have MS or cancer? In trying to come up with an answer, all I could think about were the specifics: pancreatic, colon, or skin cancer? If colon or skin cancer, with or without metastases? What type of MS, relapsing-remitting or chronic progressive? I thought on balance I'd rather have either a cancer curable by surgery or benign MS that didn't progress. But what I told the reporter was that it was an impossible question to answer.

The public relations departments of the hospitals were not the only ones who were bickering. There was some dispute among the scientists about the order of authorship on the scientific paper. There are two levels of decisions regarding authorship on a scientific paper. First, is your name on the paper or not? Then, where is your name in the list of authors? Journals and medical schools have tried to establish guidelines for whose name should or should not be on a paper. Should your name be on a paper if you had an idea but didn't do the lab work? What if you entered patients into a clinical trial but weren't involved in designing or analyzing the trial? What if you provided little else but financial support for the work? The *New England Journal of Medicine* asks that if more than twelve authors from a multicenter clinical trial or eight authors from a single-institution clinical trial are on a paper, each author sign a form affirming that he or she has complied with criteria for authorship established by an international committee of medical journal editors. Authorship attributed only to a group (such as the Multiple Sclerosis Study Group) is not acceptable; at least one person's name must accompany the group's name. As I said in the press conference, although we work collectively in science, there must ultimately be individual accountability. Most laboratory-based scientific papers in biology and medicine have between four and six authors; clinical studies have more. The number of authors has increased over the years as science has become more complex and requires larger collaborative efforts. (And authorship in other scientific disciplines [especially physics] is handled differently. A recent article in *Physical Review Letters,* entitled "Search for neutral supersymmetic Higgs bosons in pp collisions," listed 456 authors in alphabetical order.)

In biomedical publishing, authors are not listed in alphabetical order; the most important places on a manuscript are first and last. The first author is the one primarily responsible for the work, and the last author is the leader of the scientific team and is also usually responsible for obtaining funding for the research. Friendships and collaborations have been destroyed by conflicts over where people should be listed on a paper. I have learned over the years that this can often be avoided by deciding before the research starts the place of each author on the manuscript, so that people adjust their expectations. In the instance of our *New England Journal of Medicine* article, there were a total of seven authors on the paper, and there was no question that Steve would be the first author and I would be the last author. But an intense battle ensued about authorship in the middle, the least prized positions. When it was finally resolved, some authors even stopped socializing and playing squash with each other, though as time went on and people moved on to other projects it seemed less important and the hard feelings were forgotten.

In the almost two decades since I first appeared before the press, it has become more commonplace for scientists to be on television, though most scientists still feel a negative twinge when they see a fellow scientist on TV, especially one working in the same field. One of the problems is that the press can never report all the scientific nuances of the work. "Breakthrough" research invariably needs to be repeated and tested by other scientists before it is verified. Patients want to know about cures, about breakthroughs, about miracles. Furthermore, people respond strongly to individual stories about sick people being healed, even though an individual case cannot be representative of the whole. When a patients asks me about what causes multiple sclerosis and when we will have a cure, it can take up to a half hour to explain all the nuances, after which the response is often "It's really complicated, isn't it?" One of my patients has told me that she has learned to not pay attention to what is reported in the press about MS; there are too many false starts, too many disclaimers, too much false hope. "I know progress is being made," she said to me once, "but it is at a different pace than what is reported on television." Nonetheless, part of most MS-related appointments involves looking at clippings from newspapers or stories from the Internet that the patient has brought.

The lights from the press are unavoidable when a paper on MS is published in a prestigious journal, but they fade quickly and one is left with the reality of what was published and what it truly means. More important than the editorial on the Ingelfinger rule was the one written by Dale McFarlin as he analyzed the results of the two MS studies. McFarlin addressed issues that would apply to MS for the next twenty years. At the time of the editorial McFarlin was one of the senior research figures in multiple sclerosis, a much-revered individual who set up the neuroimmunology branch at the National Institutes of Health. His untimely death took from the field one of its major figures and rational voices. In his editorial there were pearls of wisdom regarding MS that hold to this day.

He began by stating that no form of treatment had been demonstrated to alter the long-term course of multiple sclerosis. It was not because of lack of effort, as hundreds of treatments had been reported since Charcot's original description of the disease, and McFarlin pointed out that a synopsis of these treatments was contained in a recent book published by the National Multiple Sclerosis Society called *Therapeutic Claims.* He then quoted Dr. Richard Masland, a neurology professor and former director of what was then known as the National Institute of Neurological and Communicative Disorders and Stroke, who commented in 1969 after the study of ACTH for MS, "There are few areas of scientific inquiry which have spawned more inadequate studies and unwarranted recommendation than that of the therapy of multiple sclerosis. The history of this disorder is one of a long and continuing series of false claims of a cure for this disease."

Therapeutic claims in MS are drenched in cynicism even to this day, and McFarlin listed six reasons therapeutic claims have fostered controversy: (1) the cause is unknown, (2) there is no good laboratory marker for the disease, (3) MS varies among patients, (4) symptoms are exacerbated by fever, activity, and emotional components, (5) placebo effects occur in clinical trials, and (6) the disease is variable and doesn't shorten life in most patients, so treatments associated with major side effects are not given early in the course of the disease, when such treatments might be most effective. Advances in our understanding of MS over the past twenty years have addressed some of these issues and are why we now have FDA-approved drugs for treatment of MS. We better understand immune mechanisms in the disease, we have MRI imaging to measure response to therapy, and we understand clinical subtypes better. But we have not yet taken the bold step of treating aggressively with drugs that have potential side effects early in the course of the disease, when the drugs might be most useful.

McFarlin went on to review the two MS studies in that issue. In discussing the hyperbaric oxygen study, he suggested more extensive investigation to include crossover design and electrophysiological studies to understand why a benefit was observed. Regarding our study of immunosuppression, he stated that the differences were convincing. He wrote that unfortunately the improvement was transient, and during the second or third year, nearly 70 percent of the patients who had initially improved had shown evidence of deterioration. He cautioned, as we did in our paper, that not all patients should be treated with these agents and that perhaps our findings could be improved by multiple courses or maintenance therapy. He noted that the positive effects we observed implied the immune system was involved in the disease, and closed by writing that he hoped that "neither of these reports will ultimately contribute to the long list of treatments for MS that have been unconfirmed and that the thousands of patients with this disorder will not again be disappointed."

As it turned out, the positive results of hyperbaric oxygen therapy could not be replicated by other double-blind studies, and so this treatment for MS was abandoned. The use of cyclophosphamide had its own stormy path, as not all investigators found positive effects, especially a study that was conducted by a Canadian group and was better controlled than ours. Nonetheless, our study was the harbinger of approval by the FDA in 2000 of another chemotherapy drug for use in treating MS, and cyclophosphamide is currently used in MS centers in the United States, Canada, and Europe in patients with active disease that has not responded to approved injectable medication.

Our report created both a stir, because we claimed in an article published in one of the most prestigious medical journals that there was a drug that could help patients with progressive MS, even if only temporarily, and a controversy, because our study design by necessity could not be perfect and the drug had side effects. In letters to the editor after publication of our paper one group criticized our results and argued that we hadn't even established temporary efficacy because our study was not performed in a blinded fashion, whereas another group reported they were using IV cyclophosphamide in MS patients and were finding positive results similar to ours. Over the course of the two decades after we treated out first patient with cyclophosphamide, there would be controversy in the medical literature about its use, a successful $1.3 million lawsuit brought by a patient in California against Blue Cross Blue Shield for denying the treatment, and finally a better understanding of why it dramatically helped some MS patients and not others.

Cyclophosphamide therapy basically became referred to as "chemotherapy" for MS. This was unfortunate, since many patients are afraid of "chemo" and didn't want to take the medicine. Furthermore, the drug has a long list of potential side effects, including bladder damage, loss of hair, infection, sterility, and an increased risk of secondary cancers, since it is an alkylating agent that affects the DNA of cells. The use of such a powerful drug in a nonmalignant disease such as multiple sclerosis raised thorny ethical questions. How much could we subject patients to, and what were the real risks?

"Are you sure that cyclophosphamide really helps MS patients?" one of my scientific colleagues asked me. We had collaborated on mouse studies together, and it was common for us to look at data from mouse experiments, question whether we were right, and then repeat the experiments to make sure there was no artifact. There was no spotlight on our mouse experiments.

"I'm sure," I told him, "otherwise we wouldn't have published it."

When the dust settled after the publication of our article, we were left with many unanswered questions and an enormous amount of work to be done. There were two major questions to answer that stemmed from our work, a clinical question and a scientific question. The clinical question was whether we could prolong or maintain the positive effect by treating patients again or giving intermittent therapy over time. This is a major issue for every MS treatment that shows efficacy—how long does it last, and can it be given over the long term? It is a question that can only be answered with time, which is frustrating to both doctor and patient. As we followed patients we had treated with cyclophosphamide, it became clear that with a single two-week treatment it took an average of a year and a half before the disease became active again. Thus we devised two other protocols, one in which we gave booster treatments following the initial therapy and another in which people could be treated every four to eight weeks as an outpatient (and didn't lose all their hair). These studies took time and were not published until 1991 as part of the Northeast Cooperative Treatment Group. Presently, cyclophosphamide is given almost exclusively as an outpatient treatment.

The second question was more basic. If cyclophosphamide truly helped patients, as we thought it did, how did it work? To our surprise, as we investigated the immunological effects of cyclophosphamide over the next twenty years, we found that it was not simply a drug that suppressed the immune system, but one that affected the immune system in a selective way that benefited MS—it decreased harmful inflammatory immune responses and reinforced parts of the immune system that had beneficial anti-inflammatory properties (points four and eighteen of the twenty-one points).

ONE OF THE reasons Houston Merritt wrote that publishing something positive on the treatment of MS could ruin a scientific career is because studying MS is full of traps that make it easy to get a wrong answer. Multiple sclerosis is unpredictable. We don't understand its cause. It expresses itself over long periods of time. McFarlin had outlined all the traps in his editorial.

"What was the major contribution of our *New England Journal of Medicine* paper?" I asked Steve twenty years later at an MS meeting in Los Angeles.

"We showed that MS indeed had an immune basis," he said, "and by suppressing the immune system, one could positively affect the course of the disease in some patients."

I agreed with him, thinking of all the people who had been helped by the treatment over the years.

The beauty of the scientific process is that there is a truth to be found. Science on one level could be considered a game. It is a question-and-answer game played by the scientist with nature. The rules are simple, but they don't necessarily favor the scientist. The scientist is allowed to ask any question, and nature must give the correct answer. No lying. The problem, of course, is asking the right question and then knowing how to interpret the answer.

As I studied the use of cyclophosphamide in MS over the next twenty years, I learned that we were both right and wrong about the effects of cyclophosphamide in the treatment of progressive MS. I had asked an important question about the treatment of MS and had gotten a correct answer. But I had also fallen into one of the traps that Houston Merritt had written about, and our publication in the *New England Journal of Medicine* in 1983 turned out to be something that haunted me on and off over the next twenty years.

9

T CELLS AND TOLERANCE

ALTHOUGH OUR RESULTS in 1983 with cyclophosphamide gave a major boost to the immune theory of MS, it was clear that cyclophosphamide was not a cure for MS; rather, it was a clue to the direction we wanted to head to find an effective MS therapy. In the 1980s work was ongoing testing other types of immune suppression, including total lymphoid radiation, other immunosuppressant drugs, and an immune modulator called copolymer-1. The viral theory also continued to be explored, most notably with the testing of interferons, which would proceed throughout the 1980s and culminate in the first approved drug for MS in 1993, even though the mechanism of action of interferons would in the end be related not to viruses but to the immune system. Investigators also continued to search for infectious agents in MS, and there were reports of a virus related to HIV and herpesviruses; however, none of these investigations bore fruit.

If MS was truly a T-cell-mediated disease, then we could treat it by specifically targeting T cells in a way that was not harmful to the patient. More was being learned about T cell biology and how T cells regulated the immune response, and technological advances were providing tools to manipulate and study T cells. Thus we embarked on three clinical experiments designed to specifically target T cells and stop the progression of MS. Although the experiments were not as successful as we had hoped, they provided insight into the workings of the immune system and MS, the principles of which were subsequently translated into therapies that have been shown to have efficacy in MS patients.

Monoclonal Antibodies

Because we believed that the effect we observed with cyclophosphamide was due to its effect on T cells, we searched for a treatment with less toxicity that would have the same effect. We turned to monoclonal antibodies, a discovery that captured the imagination of the scientific community in 1975 when Kohler and Milstein published a paper in *Nature* entitled "Continuous cultures of fused cells secreting antibody of predefined specificity," and that won them a Nobel Prize.

What are monoclonal antibodies? As discussed earlier, an antibody is a Y-shaped structure that binds specifically to its target. For example, antibodies that occur after polio vaccination bind to the poliovirus but not to measles virus. Antibodies have been used in science for many years, and before monoclonal antibody technology, antibodies were made by immunizing animals (usually rabbits) with a protein and then using the part of the animal's blood (serum) that contained many different types of antibodies. Antibodies are made by white blood cells called B cells, and each individual B cell is like a factory that produces a unique antibody targeted to only one substance, just like every person is a unique individual. The breakthrough came when Kohler and Milstein, who were investigating a different question, discovered that they could immortalize a single B cell with a transformed or cancer-like cell and then grow the fused cells indefinitely; thus they had a limitless number of identical fused B cells that all produced the same antibody—a monoclonal antibody. In essence, they had devised a way to immortalize a single antibody-producing B cell and replicate it to billions of cells. The implications of their discovery were obvious. Could one make a specific monoclonal antibody against a tumor and use it to treat cancer? Could one make a specific antibody against an unwanted type of cell in the body and eradicate only that type of cell without causing damage to the entire body? After their discovery, it was hoped that monoclonal antibodies would be "magic bullets" that could be used to treat a wide variety of diseases.

When we began our studies in MS patients in 1984, monoclonal antibodies had been used in cancer and transplantation but had never been applied to patients with an autoimmune disease such as multiple sclerosis. We postulated that the use of an antibody that killed T cells might be like cyclophosphamide and be of great benefit to MS patients.

One of the next fellows in the lab was David Hafler, a neurologist with a passion for science and the study of MS. As a college and medical school student he began his study of MS by identifying anti-MBP antibodies in the spinal fluid of MS patients. From the time he joined the lab, his scientific focus has been on T cells in multiple sclerosis, and we have worked together for almost twenty years, publishing over seventy scientific papers together. He now, like myself, has an endowed chair at Harvard for the study of multiple sclerosis and recently joined forces with Steve Hauser in the study of MS genetics.

David and I spoke to Stuart Schlossman and Ellis Reinherz at the Dana Farber Cancer Institute, who were pioneers in making monoclonal antibodies that identified different types of human T cells. David shared my enthusiasm for moving quickly to test new therapies in MS. We immediately obtained the cells that produced the monoclonal antibodies, grew them in the laboratory, purified them, tested them to make sure there were no toxic reactions in animals, put a protocol through the institutional review board, and within a year were ready to treat our first patient. Dave Dawson joined the effort and we began recruiting patients.

We did not have financial support to begin the treatment program, but we had begun to raise money for new initiatives and to build an infrastructure for studying MS. I knew then that the complete study of MS required both funding from conventional sources such as NIH and the National Multiple Sclerosis Society and private donations. In a way, it was like a special operations force in our attack on the disease. The father of a young woman with MS who helped us establish our MS program once said to me, "Howard, I never want you to be in a situation where you want to do an experiment and can't do it because you don't have enough money."

It would not be difficult to recruit patients for the monoclonal antibody study. One of the initial people we spoke to was a twenty-nine-year-old woman named Charlotte Bryant whose MS had been worsening over the past two years. We discussed with her the possibility of taking cyclophosphamide, but she was concerned because she wanted a second child and was worried about the effects of cyclophosphamide on fertility. When we told her that we would be testing a new drug that we hypothesized would be as good as cyclophosphamide, didn't affect fertility, and did not cause nausea and hair loss, she jumped at the chance.

Fresh from what we had seen with the cyclophosphamide, we hoped that we would treat Charlotte, and when we evaluated her at six months we would find that her MS had stabilized or perhaps even improved. We would measure the effect of the monoclonal antibody treatment by clinical response, as there was no MRI imaging at the time. We would, however, do a spinal tap before and after treatment to measure how much of the antibody entered the spinal fluid. Although we didn't know it then, this was a wise choice, as it would lead to an important discovery about MS.

In the end, we treated a total of twenty patients whose MS had worsened in the previous nine months, a group similar to those we treated in our cyclophosphamide study. The first antibody we tested was called anti-T12. It was targeted to a structure on the surface of all T cells. Later, we treated 8 patients with two other monoclonal antibodies: anti-T11, which also bound to all T cells, and anti-T4, which only bound to a subset of T cells, a subset we felt was more directly involved in MS.

After my experience with the press following publication of our cyclophosphamide results, I decided to keep our clinical trials with monoclonal antibodies as low-profile as possible. The trials were explained to our patients and discussed with scientific colleagues, but when the press called to ask about what was happening in MS, I shied away from providing details of our monoclonal antibody therapy. But then, in the second year of our trials, Larry Steinman published an article in *Science* reporting that monoclonal antibodies were effective in treating EAE, the animal model of MS. Reports of the success of a new treatment in the EAE model published in *Science* or *Nature* always create a stir, even though the treatment may not be effective in MS, and they always trigger a flood of calls from patients and family members. When Steinman published his paper in *Science,* everyone asked when monoclonal antibody therapy would be tried in MS. When people discovered that my group was already doing it, we had to face the press once again.

One of the patients we treated was a man named George Baker. George came from Maine and was referred to our center because of continued worsening of his MS. When we told George that the protocol called for a spinal tap to be done before and after the infusions, he told us to do as many spinal taps as we needed. "I want the experiment to work, and I want you to get as much as possible out of it," he said. In addition to the experiment working, of course, we were all hoping that the progression of his disease would be halted. He had two sons with whom he liked to play baseball. He was a true fighter.

After the anti-T12 had been prepared in a laboratory, David and I stood by George's bedside as the first infusion was being given. We didn't expect any immediate clinical effects, and there was no MRI to follow the effect of therapy. Nonetheless, we wanted to be there.

"How do you feel?" David asked.

"Fine," George said with a smile on his face. He looked relaxed and hopeful.

We stayed there for an hour, then headed back to the laboratory. A few hours later, however, we received an emergency call that George was experiencing difficulty breathing and had a small rash. We rushed back to the floor to find that he was having a mild allergic reaction to the infusion. It subsided without treatment, and George said, "You're not going to stop, are you?" The mild allergic reaction was not unexpected, and for future infusions both George and the other ten patients were given a small dose of oral prednisone to fight any allergic reactions. The prednisone worked, and we were able to treat all twelve patients with the anti-T12 monoclonal antibody.

Nine months after therapy, out of the twelve patients, eight had worsened, three were stable, and one was listed as improved. The results were nowhere near as good as what we observed with cyclophosphamide and were no better than ACTH. When we carried out laboratory assessment, we did not see any significant indicators that the antibody was suppressing the T cells. Moreover, we observed something that made giving this particular antibody problematic. The antibody we administered came from a mouse; thus it was a foreign protein for humans. We observed that two weeks after the infusions, the patients' immune systems began making antibodies to the antibodies we were infusing. In a way it was like rejecting a foreign kidney—the patients' immune system could not tolerate the foreign antibody infusion. On the positive side, we found that the antibody did bind to the T cells in the bloodstream, and apart from small allergic reactions there were no adverse side effects.

We then tested the other two monoclonal antibodies in eight patients: anti-T4 and anti-T11. This time we were careful to clear the antibody preparation of any aggregates that we believed may have contributed to reactions we observed with anti-T12, and patients were given Tylenol and an antihistamine to prevent allergic responses. They did not require any cortisone. This time the immunological results were encouraging. We saw greater suppression of the T cells, and although patients again made antibodies against the mouse antibodies we infused, they were at a lower level. The number of patients we treated was small and we did not observe clinical improvement. Later Steinman's group carried out a large-scale study with anti-T4 monoclonal antibody in MS and found some positive effects, though not enough to pursue further.

However, in the course of treating patients with the anti-T11 monoclonal antibody, we made an exciting discovery about a mechanism by which T cells cause MS. Our unexpected discovery reinforced my philosophy that a crucial part of MS research is studying the patient directly. Many discoveries in medicine come from unexpected observations, but one has to be looking in the right place. If one randomly walked through a house where there is no treasure, one could not stumble on it.

David Hafler made an important discovery because he was carrying out repeat spinal taps on MS patients being treated with a monoclonal antibody. I remember his call to me from the laboratory, where he was examining cells taken from the spinal fluid under the microscope. "I think we've discovered something," David said excitedly. I immediately joined him at the double-headed microscope as he focused on the lymphocytes taken from the spinal tap after the monoclonal antibody infusion. The lymphocytes were glowing green. "Now look at this slide," he said. On the second slide I could also see lymphocytes, but they were not glowing green.

"We labeled cells in the bloodstream with the monoclonal antibody, and it looks like the cells moved into the brain and spilled into the spinal fluid," he said with a smile. "We now have direct demonstration that in MS patients, T cells are moving from the bloodstream into the nervous system. What's more, they are doing it very rapidly," David said. "I did this spinal tap a day after the infusion."

David pulled out a piece of paper and wrote down the points that led to his conclusion. First, the monoclonal antibody that we infused into the bloodstream did not enter the spinal fluid. We knew this because even though we could find very high levels of the infused monoclonal antibody in the bloodstream, there was no free monoclonal antibody in the spinal fluid. Second, T cells in the spinal fluid from MS patients who were treated with the monoclonal antibody had the antibody on their surface. We knew this because we could visualize the monoclonal antibody that we infused stuck to the surface of the T cell using a green dye that bound the monoclonal antibody. This meant that the monoclonal antibody we infused attached to the T cell in the bloodstream and hitched a ride into the spinal fluid. In other words, in MS patients T cells from the bloodstream were continually moving into the spinal fluid (point nine of the twenty-one points). We had inadvertently labeled the T cells and then discovered them in the spinal fluid.

How did we know what we observed was not an artifact? Had the appropriate controls been done? The answer was yes. There were two controls. First, spinal fluid T cells taken prior to the treatment did not light up. Thus what we observed was related to the treatment. And second, we knew the effect was not due to monoclonal antibody that had leaked into the spinal fluid because the spinal fluid alone did not label the T cells.

The therapeutic implications of our finding became clear a few years later when studies in the EAE model showed that if one blocked movement of T cells from the bloodstream to the brain of EAE animals, they did not get paralyzed. Stopping the movement of T cells into the brain was accomplished by using a monoclonal antibody that blocked a structure on T cells that was required for the T cell to bind to blood vessels. If the T cell couldn't bind to the wall of a blood vessel, it couldn't leave the blood and enter the brain. If it couldn't leave the blood and enter the brain, it couldn't cause damage to the myelin sheath.

After positive results in the EAE model, clinical trials in MS patients were finally performed using a monoclonal antibody that bound to a structure on the surface of T cells called alpha-4-integrin. The monoclonal antibody was called natalizumab (trade name Tysabri; currently being developed by Biogen Idec and Elan). A first trial was carried out in the 1990s in which natalizumab was given to MS patients suffering from acute MS attacks, to determine if they recovered better following treatment. Unfortunately, no positive results were found. A second study was then performed in which patients with the relapsing-remitting form of MS were given monthly infusions for a total of six months and the number of attacks was measured. The trial included 216 patients and had a control group that received a placebo. These results were published in January 2003 and were clearly positive. There were far fewer attacks in the treated patients than in those receiving placebo. In addition to reducing attacks, brain inflammation was also decreased as measured by MRI imaging. There were few side effects, and there were no major problems with the body rejecting or making antibodies against the monoclonal antibody. Technology has advanced so that monoclonal antibodies are now humanized (do not have portions that are mouse in origin) and are thus tolerated much better than the monoclonal antibodies we first tested in MS patients. Tysabri had positive results in phase three testing and was FDA-approved for MS (see Chapter 12). One caveat with the Tysabri results to date is that when the treatments were stopped, relapses began once again. This means that T cells in the bloodstream of MS patients are still poised to enter the nervous system and cause damage.

With a better understanding of the immune system and the ability to make humanized monoclonal antibodies, the door has opened up for monoclonal antibody therapy for many diseases. The first monoclonal antibody received FDA approval for kidney rejection in 1986 and the second for cardiovascular disease in 1994. Thus, although it took time from Kohler and Milstein's discovery in 1975, presently there are eleven monoclonal antibodies that have received FDA approval for the treatment of conditions such as transplant rejection, lymphoma, leukemia, metastatic breast cancer, Crohn's disease, and rheumatoid arthritis. At least four hundred more monoclonal antibodies are in clinical trials. Besides Antegren, there are several monoclonal antibodies currently being tested in MS patients, each one directed against different components of the immune system: activated cells, B cells, and receptors on cells that are important in regulating an overactive immune system. Thus the discovery of monoclonal antibodies, the ability to humanize them, and our better understanding of the immune system have opened for investigation the entire area of immune manipulation in patients with multiple sclerosis.

The advance with monoclonal antibodies stems from a very basic observation made by Kohler and Millstein, who were studying B cells and had no inkling of the impact that monoclonal antibody technology would have on medicine. Even more than for therapy, monoclonal antibodies have become standard diagnostic and laboratory reagents for basic research and study of disease. They are crucial tools for the study of the immune system in MS.

T Cell Vaccination

Our search for more specific treatments of multiple sclerosis and targeting the T cell were not limited to monoclonal antibody therapy. The monoclonal antibody therapy we attempted targeted an entire class of T cells. Could we target only those T cells in MS that were causing damage to the nervous system? That in essence has always been the holy grail of MS therapy: find a treatment specific for the disease that doesn't cause side effects. Remember from the seminar in Henry Claman's laboratory that if MS is an autoimmune disease, one has to identify the T cells or antibodies causing the damage and show that if one decreases them, one can help the disease. As with virtually all immune approaches done in MS, so much stemmed from the EAE model, which serves as the best model for a T-cell-mediated disease.

A breakthrough in our understanding of T cells in the EAE model, and one that has served as the foundation for studying T cells in MS and led to our initial trials of T cell vaccination, was made by Irun Cohen at the Weizmann Institute.

What did Cohen show? Together with Avi Ben-Nun and Hartmut Wekerle, Cohen showed in animals that he could transfer EAE from one animal to another by using clones or lines of T cells specific for myelin basic protein. Although we had known before that T cells caused EAE, he directly isolated them and proved it. In a way, it was like using a virus to transfer disease. It is known that diseases caused by viruses are transmitted when the virus that infects one person is transmitted to another person. Vaccination strategies are designed to build up the immune system so it can fight against the virus. Cohen's idea was to treat the T cell with which he could transfer EAE as though it were a virus and use the T cell itself as a vaccine. In other words, if T cells specific for MBP could cause EAE, would weakening them like one does a virus and injecting them into an animal cause the animal to mount an immune response against the disease-inducing T cells and thus protect the animal from the disease?

Cohen and colleagues showed that the answer was yes. By vaccinating the animals with weakened T cells, the animals' own immune system then generated other T cells that destroyed the disease-inducing T cells. The big question was whether this approach would work in MS patients, and how it could be done. Cohen came to our laboratory to try. We didn't know which T cells to use, although a major theory was that T cells reacting against MBP could be the culprit in MS. This is the protein Salk used in his experiments. At the time Cohen was in our laboratory, it was much harder to grow T cells against MBP in humans than it is now, and it had not been conclusively shown that they were abnormal in MS patients. Thus, we reasoned that if there were T cells causing damage, they must be in the brain and we could harvest them from the spinal fluid.

David Hafler took advantage of the T cell cloning and growth techniques he had developed in the laboratory, and we grew T cells from the spinal fluid from MS patients and made them into a vaccine. The T cells were grown under sterile conditions, then weakened. A million cells were injected into the patients twice. Linda Berman was the first patient to receive a T cell vaccine. She was a dark-haired, intense woman in her thirties who wanted everything possible done to try to treat her MS but had not responded well to the cyclophosphamide. It was an exciting moment when we injected the vaccine; we even took a picture of it. We knew that it is a big step when one moves from animals to people. It creates a new fact, addresses issues of toxicity, and opens the way for others. We treated four patients. The goal of our research trial was to make sure it was safe and to see if we could find any immunological effects. For what we tried to do, our results were a success. We were able to show immunological effects in blood samples from people who had received the vaccine. Clinically, it was hard to know whether the patients improved. The effects were certainly not as dramatic as cyclophosphamide.

However, based on Cohen's pioneering experiments in animals and our initial attempts, a number of investigators began exploring the possibility of T cell vaccination in MS patients. Jeffery Raus in Belgium treated a group of patients; Zingwu Zhang, a fellow who trained with David Hafler, initiated trials in Texas. Arthur Vandenbark in Oregon took the T cell vaccination a step further by taking parts of the T cell receptor rather than the whole cell for the vaccination paradigm, and Leslie Weiner in Los Angeles and Anat Achiron at the Tel Hashomer Hospital in Israel are in the midst of a large-scale experiment using T cell vaccination. Some of the results thus far have been encouraging, although the ultimate place of T cell vaccination in MS remains unknown.

There remained a major problem with immune-specific therapy, whether it was T cell vaccination using clones against MBP or Salk trying to desensitize with MBP. Was MBP the primary target of attack by the immune system in MS? Was it only MBP, or were there other antigens, and how could one measure them? (point two of the twenty-one points) This is a key question that we addressed as we embarked on our most exciting approach to treat MS by a specific oral vaccine using the concept of oral tolerance. The work began in the mid-1980s as we struggled both to find specific and nontoxic therapies for MS and to understand what was wrong with the immune system in MS. Remember, we all have T cells in our body capable of attacking the brain and causing MS. The central question is why don't they cause disease in most people and how can they be shut off in a natural way? We believe that there is something wrong with the immune system in MS, but we don't know exactly what it is (point eight of the twenty-one points). At the center of what may be wrong with the immune system in MS—and all autoimmune diseases—is the concept of immunologic tolerance, which tries to explain how the immune system knows to attack every conceivable infectious agent we encounter in the environment but never to attack itself.

Tolerance: Why Doesn't the Body Attack Itself?

We live in a hostile environment and are always defending ourselves. The hostile environment exists on many levels: nations at war with each other, competition between people, hurricanes and tornadoes. But perhaps the most sophisticated and important defense system we have is our immune system. The immune system protects us against every possible infectious agent we confront in the environment—bacteria, viruses, parasites. Without the immune system, we could not survive. A person with AIDS dies of infection because of a compromised immune system. The immune system has been likened to an army that protects us against foreign invaders, and because there are so many different types of enemies, the immune system has developed multiple strategies to defend against enemies in the hostile environment. Many strategies are needed because there are so many infectious agents and each one is different. Unfortunately, sometimes armies encounter "friendly fire," when the army mistakenly attacks its own troops. This is what happens in autoimmune diseases such as MS.

The immune system not only acts as an army that protects us against infectious agents, but it has the amazing property of being able to distinguish between tissues in one's own body and foreign tissue. The commonest example is that of transplantation. If one transplants a kidney between two people who are not identical twins, the person who receives the transplant will reject the transplanted kidney unless medication is given to suppress the immune system. The word *tolerance* has been used in immunology to refer to the fact that the immune system does not react against its own tissue, only against foreign invaders. It "tolerates" its own tissue. It is when the immune system reacts against its own tissue that autoimmune diseases such as multiple sclerosis result. The next question is obvious. If MS is a disease in which there is a breakdown of self-tolerance, where is the defect? Unfortunately, although there are many postulates, the precise defect is not known.

If we understood the mechanisms of tolerance, this could provide a way to correct the defects of tolerance that we believe sit at the heart of MS. Although the mechanisms of tolerance are still being unraveled, a great deal is already known.

A major portion of tolerance is learned at birth. One of the first experiments to demonstrate how tolerance to one's own tissues might occur were carried out by Billingham and Owen in 1956. Remember, apart from identical twins, the immune system will attack cells or tissue from another person. Billingham and Owen carried out an experiment in goats. If one takes goat A and gives it cells from goat B, goat A's immune system will attack and destroy the cells from goat B. If goat A learns the difference between its cells and the cells of goat B in utero or at birth, what would happen then? As it turns out, Mother Nature has carried out the experiment. There is a situation called chimerism, where two different organisms (say, goat A and goat B) are joined together in utero and cells from one pass across the placenta into the other. In this instance, when goat A becomes an adult, one can inject cells from goat B and it will not reject them; it does not recognize them as being foreign.

These experiments showed that tolerance could be learned and wasn't an innate property of the organism. In the chimerism experiment the immune system of goat A was "tricked" into believing that the cells from goat B were self, not foreign. How did that happen?

The cells of the immune system that are primarily responsible for tolerance are T cells. The T cell is the most sophisticated cell of the immune system. It is the last to appear in evolution, and it is the controlling cell of the immune system. We believe that abnormal T cells cause not only multiple sclerosis but other autoimmune diseases such as juvenile diabetes and rheumatoid arthritis. As discussed before, the T cell gets its name because it develops in the thymus gland.

What happens in the thymus that creates a T cell? Remember, the primary function of the T cell is to protect the body against infectious agents, and it must be able to recognize all possible infectious agents in the environment—not a small task. The T cell does this through the T cell receptor, a structure on its surface that is like a key that fits into a lock. The T cell receptor is created by the mixing together of T cell receptor genes, which come together in a random way to create the T cell receptor. This mixing takes place in the thymus. Each T cell receptor has a different specificity. Thus, each T cell has on its surface a unique receptor or key that fits a single lock. T cells with millions of specificities are created.

Having a T cell receptor on its surface is only half of what happens when T cells are created in the thymus. Before they leave the thymus and enter the bloodstream, each T cell goes through a process called positive or negative selection. During this process, the T cell interacts with the tissue of the thymus and survives or dies depending on whether it is given a thumbs-up or thumbs-down by the thymic tissue. A thumbs-up is called positive selection; a thumbs-down is negative selection.

For many years it was believed that tolerance was caused by the negative selection of T cells that could damage one's own tissue, and MS was a failure of the thumbs-down process in which cells that could react with myelin somehow escaped. We now know this isn't the case. We all have cells in our bloodstream that were given a thumbs-up in the thymus yet are capable of causing MS (point four of the twenty-one points).

A second mechanism of tolerance relates to how the T cells are triggered. We now know that more than one signal is required to turn on a T cell: one through the T cell receptor and a second or co-stimulatory signal through other structures. The types of second signal can determine whether or not a T cell capable of attacking the brain is activated. Indeed, there are drugs currently being tested in MS that affect these second signals.

A third and perhaps the most relevant area of tolerance as applied to MS relates to regulatory or suppressor cells. The thymus not only creates T cells capable of causing MS but also creates regulatory T cells that can keep disease-inducing T cells in check. Over the past twenty years there has been controversy over the existence of regulatory cells as a mechanism of tolerance, but they are now in the mainstream of immunology, even if they are not completely understood. One of the major themes in studying immune mechanisms in MS relates to trying to understand if there is a defect in regulatory cells in MS patients and finding ways to induce (that is, enhance the action of) regulatory cells. Indeed, some of the treatments that have been shown to help MS, such as glatiramer acetate, appear to be acting by inducing T cells with regulatory properties.

As we thought about new ways to treat MS patients and were carrying out experiments with monoclonal antibodies and T cell vaccination, I stumbled upon an approach for inducing tolerance and treating MS that led me and my laboratory on a fifteen-year odyssey that continues to this day.

Oral Tolerance

My wife read in the paper that men think about sex every fifteen minutes up to the age of forty, after which they think about it every thirty minutes. We are hard-wired in this way. Scientists are hard-wired in additional ways. I think about multiple sclerosis all the time. I am always turning over the various problems associated with the disease in my mind, trying to integrate or reject the latest pieces of information about the disease gathered from a scientific meeting, a journal article, or seeing a patient. I am not alone. Most scientists will tell you they are obsessed with the problem they are studying. One of my first mentors, Bernie Fields, said to me, "You have to dream about your work. If you don't, you're no good." The obsession can change quickly if one changes fields or the nature of the scientific problem changes drastically. Scientists are quite fickle in that way. When I worked on viruses with Bernie, I read in detail articles about viruses and began dreaming about them. The instant I stopped working on viruses, I stopped dreaming about them. The problem of a disease such as MS is in some ways different from basic science, in which the answer to one question invariably creates new questions to be answered. Rather, with a disease there is a solution, an end point.

Whenever I read scientific journals, I am always scanning for new findings that may be applicable to MS. In 1985, crunched into a small seat on a plane on my way to Allentown, Pennsylvania, to visit Larry Levitt and work on our neurology handbook, I was going through the process of turning the disease over in my mind while reading the *Journal of Immunology* when an article caught my eye. The article dealt with the effect of orally administered proteins on the immune response, something called oral tolerance. I had briefly worked in this area before, but it wasn't until this particular instant that the idea crystallized in my mind. There was a scientific precedent for it that went back to the beginning of the century, the treatment would have no toxicity, and it would be as easy as feeding someone common proteins. It seemed too good to be true, but I could not find anything wrong with the idea, and it was easily testable. As soon as I returned to the lab I had my postdoctoral fellow carry out an experiment on rats.

The concept of oral tolerance is very simple. We take in many proteins through the gut, via the food we eat, and the body does not react negatively to these proteins. The gut tolerates proteins given by the oral route, since taking in proteins this way is central to our survival. This is so despite the fact that the largest portion of the immune system in the body is in the gut. The experiment was simple: feed rats myelin basic protein and see if it protects them from paralysis when they are given EAE. As a control, feed another protein, ovalbumin. The results were clear-cut. The animals that were fed myelin basic protein did not become paralyzed, whereas those that were fed ovalbumin did. The results seemed too good to be true. Could treating MS be as simple as feeding patients a brain protein and having the gut's natural immune system shut down the disease?

I was not the only one studying oral tolerance to EAE. Caroline Whitacre, a scientist at Ohio State University, had found identical results. Both our labs carried out extensive experiments to understand how oral tolerance worked in animals. Caroline found that feeding high doses of MBP paralyzed the T cells causing the damage. We found that by feeding low doses of MBP we could induce suppressor or regulatory cells, which we could then transfer from one animal to another to shut down the disease process. It was only logical that an organ as important as the gut would use more than one mechanism of tolerance to protect against reacting in a harmful way to ingested proteins.

We moved as quickly as possible to test the concept of oral tolerance in MS patients. I was joined both by Dave Dawson, who ran the pilot clinical trial of oral tolerance in MS, and by David Hafler, who performed immunological studies on the patients we treated. Hafler himself had carried out early experiments testing immune responses to myelin antigens in the gut. In the pilot trial we fed bovine myelin that was prepared by Ahmad Al-Sabbagh. Only thirty patients were treated, but there was a suggestion of positive results in the men in the study, and we found positive immunological effects in the blood of those who were given the oral myelin. The results were deemed worthy enough that they were published in *Science,* as was an initial trial of oral collagen in rheumatoid arthritis.

The translation of a new treatment for MS from the laboratory to an FDA-approved drug requires millions of dollars and industrial support. The Brigham and Women's Hospital helped us establish a biotechnology company called AutoImmune, which took on the development of oral myelin for the treatment of MS. Robert Bishop came from California to run the company. We went from thirty patients to over five hundred patients in a double-blind, placebo-controlled trial. It took almost two years to enter all the patients in the trial, and many MS centers participated. All we could do was sit and wait for the human results while we carried out additional experiments on mice and rats in the laboratory. While this was probably the most exciting and potentially far-reaching experiment I had ever done for the treatment of MS—for if the results were positive, people could take a pill for their MS that had no toxicities—at this point I sat on the sidelines and waited as the large phase three trial was conducted that, if positive, would result in a new FDA-approved drug for MS.

The theater has drama. The final minutes of a close basketball game have drama. The announcement of Academy Award winners or election results has drama. Science also has its drama, but the drama is different in that it is played out in tiny increments. Each scientific experiment asks a question, and there is a palpable excitement when the results are first viewed. However, virtually all scientific experiments do not give a definitive answer. They raise additional questions, and the answers to those questions raise even more questions. A positive experiment that provides an answer must be repeated several times to ensure that the result is not artifact. Thus, the process of unveiling scientific truths is a slow-moving one that creeps along without fanfare until suddenly it's there. Nonetheless, there are moments of high drama in science, moments when a series of experiments or multiple lines of investigation culminate in one crucial experiment, an experiment in which ten, twenty, even a hundred previous experiments are tested. The oral myelin trial was one such moment.

The answer came in one sentence as the team gathered in secrecy to hear the results. "The results of the oral myelin trial are negative." I must admit I was initially devastated by the negative results. I had held out such hope for the therapy, and the results in the animal models were so promising.

There have been many failed trials in the attempt to find an effective, safe treatment for multiple sclerosis. The oral myelin trial was so unique and had such major ramifications for the treatment of MS and other autoimmune diseases that it was written about in a book called *Human Trials: Scientists, Inventors, and Patients in the Quest for a Cure,* by Susan Quinn. Because biology deals with so many variables, especially in clinical trials, there are always variables that cannot be controlled completely, variables that cause an experiment that one thinks should be positive to be negative. In the case of the oral myelin trial, we believe that for oral tolerance to work in MS the dose of protein administered needs to be optimized and that an adjuvant needs to be given to boost the induction of regulatory T cells in the gut, an area we are now exploring in the laboratory. The development of an easily administered oral (or nasal) medication remains one of the goals of multiple sclerosis therapeutics, as all current FDA-approved drugs for MS are given by self-injection or intravenous infusion.

Despite the negative clinical results, our investigation of oral tolerance led to a critical discovery about suppressor or regulatory cells and how they work. In fact, we discovered a new class of T cells, which we named Th3 cells. Many types and subcategories of T cells have been identified based on their function, surface structures, and factors they produce. A major classification of T cells based on the factors they produce are Th1 and Th2 cells. Th1 cells produce gamma interferon and Th1 type T cells targeted to the myelin sheath cause EAE and are believed to cause multiple sclerosis. Th2 cells produce a factor called IL-4, and Th2 cells are important in immunity against parasites. We discovered that Th3 cells were different from Th1 and Th2 cells because they produced a chemical or cytokine called TGF-beta that suppressed the immune response and could suppress both Th1 and Th2 cells. Youhai Chen in the lab was able to clone Th3 cells, and Ariel Miller discovered that they worked by a concept called bystander suppression, in which regulatory T cells have a tissue specific anti-inflammatory effect in parts of the body where the fed protein is present. The concept of bystander suppression has become one of the basic concepts of how T cells of the immune system function. Furthermore, this makes it unnecessary to know which myelin protein is the target of attack in MS. Copaxone, one of the drugs approved by the FDA for the treatment of MS in 1996, induces regulatory T cells such as Th3 cells that worked by bystander suppression in the brain. (point seventeen of the twenty-one points)

But not all trials in MS fail. In my office, one of the signs on the wall is a quote from Winston Churchill, who is said to have used it as the shortest commencement speech in history: "Never, ever, ever, quit." As we will see, enormous perseverance was required to bring the first drugs to be approved by the FDA for multiple sclerosis to market.

10

A DRUG FOR MS, FINALLY

THE WORD *breakthrough* has almost become a cliché in medicine. I have had MS patients ask on more than one occasion what this week's breakthrough in MS was, and I once saw advertising for a hospital that read "Another day, another breakthrough." Nonetheless, there are moments when breakthroughs do occur, when a barrier is broken and we truly enter new territory. Such a breakthrough occurred when an FDA advisory committee recommended approval of interferon beta 1b for the treatment of relapsing-remitting multiple sclerosis. It was the first FDA-approved drug for MS, and after two other interferons were approved by the FDA, the story of MS could be divided into the pre- and postinterferon eras. Shortly thereafter, two additional drugs, a synthetic protein molecule and a chemotherapy drug, were approved by the FDA for MS. Thus, MS changed from an untreatable disease to one in which the physician and patient had multiple treatment options. The treatments were not cures, but a major step on the road to curing MS had occurred. The moments that marked the approval of each of these drugs were advisory board hearings at the FDA. Thus, an important part of the story of MS is the story of these FDA hearings and the events that led up to them.

On March 19, 1993, an FDA advisory committee recommended

approval of a new drug for multiple sclerosis: interferon beta-1b (trade name Betaseron). Evidence of the drug's safety and effectiveness was presented to the FDA advisory committee by the drug's maker, a California company called Berlex. Betaseron was officially licensed by the FDA in July 1993 for the treatment of relapsing-remitting multiple sclerosis. A little over two years later, on December 4, 1995, a second interferon, interferon beta-1a (trade name Avonex), was presented to an FDA advisory committee by a Boston company called Biogen and was also subsequently approved for relapsing-remitting MS by the FDA. In 1997, a different form of interferon beta-1a, manufactured by Serono, a Swiss company, and sold under the name Rebif, became available in Europe; it was approved in the United States in 2002. The approval of Betaseron, the first FDA-approved drug for MS, was a major milestone in the search to find a treatment and ultimately a cure for multiple sclerosis. Multiple sclerosis was classified as an "orphan disease" (those considered to be relatively rare) by the FDA based on there being less than 200,000 people in the United States with the relapsing-remitting form of MS. The approximate number of MS patients in the United States is 400,000 (some feel it is even higher), with a total of over 1 million worldwide. With the approval of Betaseron, pharmaceutical and biotechnology companies realized that large amounts of money could be made on an MS drug. Betaseron was priced at approximately $10,000 per year per patient. Thus, every ten thousand patients on Betaseron generates $100 million in revenue. With FDA approval of interferons, the study of MS would never be the same.

In 1980, when we began planning our trials of cyclophosphamide, our hypothesis was that MS was an autoimmune disease and that treatments to suppress the immune system would be of benefit. The other major theory was that MS was caused in some way by a viral infection. This line of reasoning led to testing interferons as a treatment for MS. The interferon story in the United States began virtually at the same time by two independent groups that followed different paths. Each path ultimately led to the approval of an interferon drug by the FDA.

The first path was begun by Byron Waksman at the National Multiple Sclerosis Society in New York, and the second path by Larry Jacobs, a neurologist in Buffalo. I will begin the story of the interferons with Byron Waksman.

Waksman is a major figure in the story of MS because he was head of the research program at the National Multiple Sclerosis Society from 1980 to 1990. Before that he was a faculty member at Yale, where he did seminal work on basic questions of immunology, EAE, and autoimmunity. (Barry Arnason in his early days was a fellow of Byron when he discovered the role of the thymus in EAE.) When Byron finished his tenure at the National Multiple Sclerosis Society, we gave him a position at our Center for Neurologic Diseases in Boston as a senior research fellow. He spends a few days each month in our labs and offers his perspective to young scientists who didn't read or remember the early papers. Laboratories and science have their lineages, and one could say that Byron is the grandfather, Barry Arnason a son, and I the grandson. Of course it depends on when and where you start counting, and I like to think of myself as a father with my own scientific children and grandchildren. Byron himself comes from an illustrious lineage. His father, Selman Waksman, won the Nobel Prize for his work on streptomycin in 1952.

I sat with Byron at his summer house on Woods Hole on Cape Cod, where his father used to live and where Byron and his family now spend their summers. It has become a common summer pilgrimage for me to drive to Woods Hole, sit with Byron, and talk about MS. In fact, it was a delegation led by Norman Kohn that met at Woods Hole with Byron to convince him to take on the job of research director of the National Multiple Sclerosis Society. When Byron began at the society, one of the first things he did was to organize workshops that critically discussed the many aspects of MS, and one of the first workshops he organized was one on interferons. The society later committed close to a million dollars for the study of interferons in MS.

Interferons are substances produced by cells that slow down or block the growth of viruses. Because they "interfered" with the growth of viruses, they were called "interferons," a simple and nonscientific name. The interferons were discovered in 1957 by Isaacs and Lindenmann, who detected that a substance produced by mammalian cells had antiviral properties. As more research was carried out, different types of interferon were discovered, and human interferons are now broadly classified as type 1 or type 2 depending on their biological properties and the cell surface receptors they use. Both type 1 and type 2 interferons were ultimately tested in MS because of their anti-viral properties. Interestingly, type 1 interferons made MS better, type 2 interferons made MS worse.

"What made you so interested in the interferons for MS?" I asked Byron as we sat on his lawn in Woods Hole overlooking the harbor.

"Two reasons," he said. "First, if MS is caused by a virus, treating with an interferon may suppress viral replication and help MS. Second, at that time there were publications suggesting that MS patients had a defective immune response against viruses and could not make normal amounts of interferon."

I was well aware of this potential immunological defect in MS related to viruses. My first independent grant to study the immune response in MS, from the Kroc Foundation, was to study natural killer cells, which are involved in fighting off viruses, and I later was awarded an endowed chair to study MS by the Kroc Foundation. The Kroc Foundation was established by the heirs to the McDonald's hamburger chain fortune because several Kroc family members were afflicted with autoimmune diseases, including multiple sclerosis and rheumatoid arthritis. I remember exactly where I stood in Bernie Fields' lab when I called the Kroc Foundation offices to find out if my very first independent grant to study MS would be funded. I held onto the receiver with a pounding heart and sweaty palms, waiting for the answer. That hasn't changed dramatically. I felt nervous recently opening an envelope to find if my NIH grant for the study of MS, now in its sixteenth year, would be refunded. Some seasoned performers tell of nervousness no matter how many times they have been on the stage.

One of the cell types that produce interferons are the white blood cells of the immune system. Andrew Neighbour, working at the Albert Einstein College of Medicine in New York, had reported that if one stimulated white blood cells from MS patients, the cells were defective in their production of interferon. With the grant from the Kroc Foundation, Steve Hauser and I published a paper stating that there was decreased natural killer cell activity in MS. At that time it was believed that natural killer cells were needed to fight off infections associated with viruses. Nonetheless, the overwhelming rationale for testing interferons in MS was the belief that viruses were at the root of MS and interferons could affect viruses. Equally as important, there was no treatment for MS, and as physicians watched patients become disabled before their eyes, they were always desperately looking for something new.

"We decided it was time to determine if interferons could help patients with multiple sclerosis," Byron said, "and the only way was to test it directly in MS patients." I've often had similar thoughts. One can do as many experiments as one wants in mice or in a test tube, but ultimately one has to confront the disease directly by testing MS patients.

"What about testing it in the EAE model first?" I asked.

"Interferons are unique for each species," Byron said. "Thus it was problematic to test human interferons in the animal model of MS."

Byron contacted Thomas Merrigan in California, who had done work on interferons in cancer. Merrigan said he would be willing to help with a trial but needed a neurologist and two MS centers to carry out the trial, since his only experience was in treating hepatitis. The meeting was held in a small conference room at the San Francisco airport. Merrigan was joined by Ken Johnson, head of neurology at the Veterans Administration hospital in San Francisco, and Michael Oldstone, an expert in the study of viruses and the immune system at the Scripps Institute in La Jolla.

"There was little to base the trial on," Byron said. "There had been two prior trials of interferons in MS—one by Ververken, who gave interferon to three patients with progressive disease for two weeks, and another by Fog, who treated six people with chronic progressive disease for fifteen months. Neither of them found any benefit. We had to decide how to give the interferon, what dose to give, how often to give it, which patients to treat, and what would be the trial design." Techniques had been developed in 1978 for the large-scale purification of human interferons from cells, but it was expensive. The National Multiple Sclerosis Society paid the Finnish Red Cross close to $1 million for the interferon to be used in the trial, and even at that cost, only a limited number of patients could be treated. Assuming that interferon indeed could help MS, all that was needed now was to design a trial that would ask the right question, like distinguishing between myself and the marathoner Bill Rodgers, and thus provide positive evidence for further studies. The stakes were high. If nothing was found, it would be hard to raise another million dollars.

Ververken, a European who was one of the first to test interferons in MS, argued that the interferon must be injected into the spinal fluid because that's where the MS was; if injected into the blood, the interferon could not cross the blood-brain barrier and gain access to the brain. However, the group Waksman assembled chose the subcutaneous route because they believed that the disease process in MS was not localized only in the brain and spinal cord but was systemic and involved the immune system throughout the body, (point fourteen of the twenty-one points). They wanted to affect the immune system, and they also hoped the interferon would enter the nervous system because the MS had already caused breaches in the blood-brain barrier.

Who to treat? They decided to treat patients with both relapsing-remitting and progressive disease, and they decided to treat continuously rather than just at the time of an attack because otherwise they were not sure they could be certain when the disease was active. How much to give? Under the assumption that more was better, the highest dose possible near the maximal tolerated level was chosen (5 million units of natural interferon alpha). How many patients to treat and by what design? The design of the trial was determined in large part by the scarcity and cost of the interferon. It was decided to treat a total of twenty-four patients, twelve at the University of California, San Francisco, and twelve at the Scripps Clinic in La Jolla, and to use a crossover design. Each patient was treated for six months with either daily injections of subcutaneous interferon or placebo and then taken off treatment for six months. After six months without treatment, those who were initially given placebo were crossed over and given interferon for six months, and those initially given interferon were crossed over and given placebo for six months. This design ensured that all patients could be evaluated for treatment. Robert Knobler, a neurology fellow in Mike Oldstone's laboratory, and Hill Panitch, a neurologist at the San Francisco VA hospital, took over conduct of the trial.

Although it was designed to be a double-blind trial, it was not difficult to tell who was treated, as there were side effects, including flulike symptoms, fever, and injection site reactions in patients receiving interferons. Remarkably, all of the patients finished the trial even though they felt sick for six months. Ken Johnson told me that this was a testimony to the determination of MS patients and their commitment to help find a therapy for their disease. Of the twenty-four patients treated, fifteen were relapsing-remitting and nine had a progressive component to their illness. The results? Although there were fewer attacks in the interferon-treated group and in the fifteen patients with the purely relapsing form of the disease, the overall results did not reach statistical significance ($p = 0.08$). Thus, rather than black and white, the answer was gray, though Hillel Panitch, one of the investigators who also participated in further studies of interferon, felt the treated patients did better in follow-up. There were changes in the immune system with treatment, but the changes did not link to clinical responses. The results would probably have been statistically significant if more patients had been treated and if the study had focused solely on relapsing-remitting patients (the Bill Rodgers marathon experiment). Nonetheless, it was a crucial start, and the same group would get another chance to test interferons in MS when a recombinant form of interferon became available. They found that interferon could be given systemically, and it may actually have worked in relapsing-remitting patients. As it would turn out over a decade later, Byron Waksman's instincts that interferons might be of benefit in MS would be proven correct.

AT VIRTUALLY THE same time that the meeting using systemic interferon was taking place in a small conference room in the San Francisco airport, another crucial step in using interferons to treat MS was being taken at the Roswell Park Cancer Institute near Buffalo and was led by neurologist Larry Jacobs.

In 1980, one of the major theories of the cause of multiple sclerosis was that there was a virus in the brains of MS patients, a virus that had not yet been identified. Even though viruses isolated from MS patients, such as the Carp agent, could not be replicated, the search for viruses continued. There was still no explanation for the elevated levels of gamma globulin in the spinal fluid, and many felt it was a reaction to a virus. If there was a virus in the brain and interferons could block the growth of viruses in general, one didn't necessarily need to know what the "MS virus" was in order for treatment with interferon to be effective. However, one would have to get the interferon into the brain for it to work, and the primary way to do this would be to inject the interferon directly into the spinal fluid.

Researchers at Roswell Park had experience with giving beta interferon into the spinal fluid of children who had leukemia that had spread to the nervous system. Beta interferon was also given into the spinal fluid of neonates with herpes encephalitis and adults with leukemia that had spread to the coverings (meninges) of the brain. It turned out that investigators at Roswell Park had extra interferon left over from their trials, and they approached Larry Jacobs to test it in a neurological disease that might be related to a virus. It was initially decided to test it in amyotrophic lateral sclerosis (ALS, or Lou Gehrig's disease), but when they couldn't find enough patients with ALS they decided to test it in MS patients. If MS was caused by a virus in the brain, Jacobs and colleagues reasoned, they needed to get the interferon into the brain. The experimental design they chose was both bold and flawed. The interferon would be given to MS patients via multiple spinal taps, and these patients would be compared to untreated control MS patients.

The interferon given to MS patients was made at Roswell Park by purifying the interferon from a human fibroblast cell line, adding human albumin, and freeze-drying the material. The interferon was given twice a week for four weeks and then once a month for the next five months. At each treatment the spinal fluid was taken for analysis, and a final spinal tap was done one year after the study. It was a bold experiment, both for the physicians and for the patients, who agreed to undergo thirteen spinal taps per patient over a six-month period. Twenty patients were treated, twelve with relapsing-remitting MS and eight with progressive MS in an open fashion in which the treated patients were compared to untreated control patients.

The results were published in the journal *Science* in 1981. That in itself was unusual. *Science* and *Nature* are the two premier weekly scientific journals. When the human genome was finally mapped in 2001 by two competing groups, one group published their results in *Science* and the other published in *Nature*. *Science* rarely publishes the results of clinical trials. Presumably *Science* published the Jacobs paper because the results implicated a virus in MS and because MS was a crippling disease in need of a breakthrough as to its cause. What were the results? Jacobs reported a significant decrease in the exacerbation rates of those who received the beta interferon compared to their exacerbation rates prior to being treated. Before treatment patients had an average of 2.5 attacks per year, and after treatment it was 1.2 per year. There was also a trend suggesting that more treated patients than controls were improved or unchanged. Five of the treated patients and one of the control patients improved, whereas five of the controls and only one of the interferon recipients had deteriorated. Those who improved were those who had had relapsing-remitting disease for a shorter duration and had less disability.

Although the results were provocative, it was clear to the neurological community that the Jacobs study was severely flawed. One year later, at a 1982 conference in Rochester, New York, on clinical trials where we presented initial results of our work on cyclophosphamide and plasma exchange, Jacobs presented his results on beta interferon injected into the spinal fluid and was criticized for not doing an appropriately controlled trial and for drawing conclusions from too few patients. Furthermore, it was known that exacerbation rates went down with time irrespective of treatment. In his defense, Jacobs pointed out that patients had developed low-grade fevers and a reaction to the injected interferon in the spinal fluid and that they could not give a foreign protein into the spinal fluid of controls, as it was not ethical. He also argued that the possibility of a placebo effect was minimized by the consistency of response and the length of the period of observation. He also argued that the results were consistent with the idea that there was a virus in the central nervous system that acted as a trigger for the repeated exacerbations of MS and that this trigger was suppressed or eradicated by the antiviral effects of the interferon.

Nonetheless, Dr. Jacobs knew that a better controlled trial with larger numbers of patients was needed and he presented the plan for the next trial to address his critics and establish the validity of his findings. It would be a multicenter study involving 40 patients per group and would enroll only patients with high exacerbation rates, not patients that were only progressive. When the trial was actually performed, it involved sixty-nine patients. It was a randomized, double-blind, placebo-controlled trial. The patients were given low doses of a drug called indomethacin to reduce the toxic side effects of the beta interferon. The placebo group was given a sham spinal tap (analogous to the sham plasma exchange we performed in our study of plasma exchange). The sham spinal tap was done by anesthetizing the skin over the spine and sticking in a needle but not injecting anything. Thus patients could not tell whether they were being treated or not. One physician was assigned to examine and evaluate the patients without knowing their treatment. Thus the trial was the first use of a blinded physician, and the results of the trial were published in 1986 in the British journal *Lancet*.

The results? Positive! There was a statistically significant decrease in attacks in the treated group as compared to the controls ($p = 0.04$). It appeared that what Jacobs had observed in his uncontrolled trial was real. Jacobs called Rick Rudick, a neurologist at Rochester who now is director of the Mellen Center for Multiple Sclerosis in Cleveland.

"We may have just cured MS," he told Rudick excitedly.

"It's not the end," Rudick shot back. "It's just the beginning."

Jacobs' reaction reminded me of my reaction when we reviewed the results of our cyclophosphamide trial. What made results such as these so exciting to us is that the scientific community didn't know what caused MS and no treatment had been shown in a controlled way to help MS. When we got positive results in a trial, we thought that it was only a matter of time before we would be able to turn a partially effective treatment into one that was completely effective.

Nonetheless, Jacobs would find the road a difficult one before he would be able to carry out the much larger trial needed to bring interferon therapy to all MS patients. Furthermore, the trial would be markedly different from what he imagined when he obtained positive results by injecting interferon into the spinal fluid.

One of the major problems with the use of interferons for the treatment of MS was that the interferons had to be extracted from human cells. It was a laborious and expensive process and had the potential of viral contamination. The National Multiple Sclerosis Society had to pay close to a million dollars to do a trial in a small number of patients, and the only reason Jacobs was able to test interferon in MS is that there was leftover interferon at Roswell Park from treating other conditions. The situation changed with advances in molecular biology and recombinant DNA technology that made it possible to produce large quantities of pure proteins. There was a big stir when recombinant DNA technology first became a possibility; it happened when I was a fellow in Bernie Fields' lab, and Bernie served on committees to draft guidelines for the use of the technique. Because the technique was so new, many were worried about its safety. In fact, the city of Cambridge, Massachusetts (home to a company that uses this technology), outlawed recombinant DNA investigation until guidelines could be drafted.

The currently used interferons in MS are all made by recombinant DNA technology, and it was the discovery and development of this technology that sparked the development of biotechnology. Genentech was one of the first biotech companies and one of the more successful ones, developing a number of recombinant proteins for human disease, including human insulin and human growth hormone. Other companies were producing recombinant forms of interferon and began testing them in MS. Schering-Plough, a major pharmaceutical company in Kenilworth, New Jersey, had a recombinant alpha interferon that was tested beginning in 1982 by David Camenga and Ken Johnson, who by then had moved to Maryland from San Francisco. A two-center, double-blind, placebo-controlled trial of recombinant alpha interferon was carried out in MS using the Schering-Plough preparation. The alpha interferon was given subcutaneously three times a week, but no effect was seen. Why there was a discrepancy between the natural alpha interferon trial that Waksman organized and the recombinant alpha interferon trial of Schering-Plough is not known. Was it the dose or the preparation? The dose of the recombinant chosen was lower than that of the natural interferon because of the severe side effects of the natural interferons.

In 1984 Biogen, a newly formed biotech company based in Cambridge, Massachusetts, had also manufactured a recombinant form of interferon and was anxious to test it in MS. Their interferon was called Immuneron, and John Schneider from Biogen called Hillel Panitch in Maryland. Hill was one of the principal investigators in the initial alpha interferon trials with Ken Johnson that Byron Waksman organized. Would he and Ken Johnson be willing to test Immuneron in MS? However, there was a major caveat in testing Biogen's Immuneron. It wasn't an alpha or beta interferon; it was a gamma interferon. Gamma interferon had different immune properties than the other interferons; it could stimulate the immune system.

"We felt it was an important new interferon to try," Hill told me at a neurology meeting in Denver. "We were interested in the antiviral properties of gamma interferon."

The trial of gamma interferon has become a classic MS trial, the results of which form the underpinning of one of the basic principles of the immune basis of the disease. It was a trial that some felt should never have been done, and gamma interferon will never again be given to MS patients. Why? Unlike alpha and beta interferon, giving gamma interferon to MS patients made MS patients worse by triggering new MS attacks.

To understand what happened, we have to go back to the early 1980s, when interest in interferon therapy was continuing to grow even though an understanding of how the interferons worked and the rationale for their use was still unclear. The rationale for testing gamma interferon (also called immune interferon) was similar to that for beta interferon. It had antiviral properties, and it had the possibility to correct a defect that some felt existed in MS patients: deficient production of gamma interferon by blood cells. Nonetheless, there was serious concern before the study began about treating with gamma interferon, because gamma interferon could activate the immune system in a way that theoretically could be harmful to MS patients. At scientific meetings where the plan to treat MS patients with gamma interferon was presented, some scientists warned that it could make MS patients worse. Ken Johnson told me that Hartmut Wekerle, now head of the Max Planck Institute of Neurobiology outside Munich, warned that gamma interferon would make MS patients worse, and at an American Academy of Neurology meeting I attended, Larry Steinman expressed the same concern.

Nonetheless, no one really knew what caused MS, and although there were some encouraging results from the alpha and beta interferon trials, not all interferon trials were positive. So many people had been wrong about predicting what would or would not work in MS. Thus Hill Panitch and Ken Johnson embarked on a study of gamma interferon.

It was a pilot dosing trial without a control group in which eighteen patients with relapsing-remitting multiple sclerosis were treated with one of three doses of gamma interferon by intravenous infusion twice a week for four weeks. What happened? The gamma interferon made the MS worse. With treatment, almost 40 percent of the patients had an attack, and the severity of the attack tended to be greater in those who received a higher dose. The annualized attack rate prior to treatment was 1.42 attacks per year, and it rose to 4.67 attacks per year after treatment. The p value was less than 0.01—highly significant.

Even though there was no control group, the fact that attack rates went up was felt to be of major biological significance. In most MS trials the attack rate tends to go down, both because of the natural tendency for attacks to go down with time and because of placebo effects. Fortunately, in the gamma interferon case all patients recovered from their attacks, and in the following couple of years, the attack rate returned to 1.05 attacks per year. Most attacks involved signs or symptoms patients had experienced previously, there was no increase in disability, and there was no evidence that fever or other side effects from the gamma interferon triggered the attacks.

"It wasn't a blinded or controlled study," Hill said, "but within six months it was clear to all of us that the treatment had triggered exacerbations, and we stopped the trial." The gamma interferon results were published in 1987.

Why did gamma interferon make MS worse? In addition to its antiviral effects, for which it was postulated to help MS, gamma interferon might have made MS worse by activating myelin-targeted T cells, which were felt by many to be the cells that caused MS. The drug did this by increasing expsession of the MHC molecule on the surface of monocytes, which makes it easier for the monocytes to trigger the T cells. This is why people questioned giving gamma interferon to MS patients. Indeed, point one of the twenty-one points is that MS is caused by T cells that are targeted to myelin in the brain, where they initiate damage by producing gamma interferon. Indeed, Panitch's group reported increased reactivity to the brain protein MBP in patients treated with gamma interferon. Furthermore, a major theme that runs through the twenty-one points relates to the Th1 vs. Th2 paradigm of MS. Th1 T cell responses, characterized by the production of gamma interferon, are bad for MS. Th2 T cell responses, characterized by the production of IL-4 and IL-10, are good for MS. The Th1 vs. Th2 paradigm was popularized in a paper by Mossman and Kaufman published in 1984, just at the time Panitch was giving gamma interferon to MS patients. Later, in the 1990s, we discovered that cyclophosphamide acted in MS not just by suppressing the immune system but by decreasing gamma interferon (Th1 responses) and increasing Th2 and Th3 responses.

Another reason gamma interferon was tested in MS is that there were reports that immune cells from MS patients were defective in production of gamma interferon and gamma interferon was necessary to fight off viral infections. This did not turn out to be the case. Indeed, in a carefully done study on the patients he treated and on untreated patients, Panitch reported that MS patients actually produced *increased* amounts of gamma interferon compared to controls. This has been reported by many investigators, and we have found that increased levels of gamma interferon in the bloodstream of MS patients are associated with increased disease activity as measured by MRI. Thus, rather than giving gamma interferon, one wanted a treatment that *decreased* gamma interferon. The triggering of MS attacks by giving gamma interferon was consistent with an observation by William Sibley in 1985 that MS attacks are often related to common viral infections, which are known to activate the immune system to produce more gamma interferon.

Panitch's paper has become a classic, because it showed that attacks of MS were caused by stimulating the immune system and Th1 responses. His results supported the hypothesis that MS was a systemic disease that could be triggered in the blood and then lead to an attack on the brain. Since interferons do not cross from the blood into the brain, the triggering of MS attacks by the giving of gamma interferon into the bloodstream emphasized the close connection between the immune system outside of the brain and what happens in the brain itself. This indeed is point fourteen of the twenty-one points, namely that in MS there is movement of cells from the bloodstream into the brain. As discussed before, clinical trials are now in progress using compounds that stop the movement of T cells from the blood into the nervous system.

Panitch and his group ended their publication by declaring that the use of gamma interferon was contraindicated in MS. More important, they went further and stated that compounds that inhibit gamma interferon production or its effect on the immune cells could help MS. They also argued that agents that counteracted interferon could be given systemically and that administration of interferon into the nervous system (as by spinal tap) was not needed. Indeed, beta interferon, the first class of drugs approved by the FDA to treat MS, actually caused a decrease in gamma interferon. Ironically, the gamma interferon tested by Panitch, called Immuneron, was provided by Biogen, the Cambridge biotech company that was second to the market with a beta-interferon drug for MS.

No matter what the theory about MS or any other disease, the proof is in the successful application of the theory to the treatment of the disease. This is especially true when one uses a treatment that works in animal models as a basis for testing the treatment on people. The transition does not always work. Nonetheless, a great deal was learned from the treatment of MS by gamma interferon. Whenever one studies the disease directly, there is always the chance to learn.

WHILE THE GAMMA interferon trial was in progress, the study of beta interferon, the first drug to be approved for MS, was gaining momentum. Larry Jacobs had finished his trial of natural interferon given into the spinal fluid and was looking for a source of recombinant beta interferon to test in MS patients. He would ultimately be successful, and his efforts would lead to the second interferon to be approved for MS by the FDA. After the gamma interferon trial, Ken Johnson and Hill Panitch would get a second chance to test a recombinant beta interferon in MS. They would build on the experience they had gained in testing the natural alpha interferon that the National Multiple Sclerosis Society had purchased from the Finnish Red Cross and the recombinant gamma interferon provided by Biogen. The studies would lead to the approval of a beta interferon drug by the FDA, called Betaseron. The successful development of Betaseron involved an oil company that briefly tinkered in biotechnology, and a young oncologist named Steve Marcus, without whom Betaseron would have never existed and whose name is virtually unknown to the neurological community or to MS patients.

In 1980 the oil company Shell had decided to get into biotechnology and set up a company called Triton Biosciences, headed by Dick Love, who was a chemical engineer. At that time, as Steve Marcus told me, "interferons were hot molecules." The first recombinant alpha interferon had hit the market for hairy cell leukemia. Interferon was being tested in many cancers and was on the cover of *Newsweek* and *Time*. Schering-Plough and Roche had versions of recombinant alpha interferon, and Shell felt that if it could make a stable recombinant beta interferon, it would have control of the field. So it initiated a joint venture with Cetus Corporation. If Cetus could successfully make the recombinant beta interferon, Triton could do the testing. It didn't take long. By 1983 Cetus had successfully made recombinant beta interferon, and in November of that year beta interferon was put into patients with advanced cancer.

Enter Steve Marcus, a thirty-one-year-old oncologist trained at the University of California, San Francisco, who was hired by Triton in March 1985. Marcus' first assignment was to test in patients with brain tumors the recombinant beta interferon that Cetus had made. Marcus came to know Robert Knobler, who was involved in the alpha interferon trials set up by Byron Waksman and had treated a number of patients with brain tumors. In the course of those conversations, Marcus heard about alpha and beta interferon trials in MS, and the possibility of trying beta interferon in MS caught his attention. On a site visit to Thomas Jefferson University Hospital in Philadelphia, he met with Robert Knobler again and discussed the use of interferons for MS. At this time there were reports of Jacobs' trial and interferon helping MS patients when given via spinal tap. Like oncology drugs, interferons made people sick and people felt terrible on large oncology doses of interferons, but this didn't bother Marcus, who himself was an oncologist. Marcus was at his first job in biotechnology and wanted to develop something that was his.

After careful scrutiny of all that had been done to date in the interferons, it crystallized in Marcus' mind that there indeed was something to the interferons in MS, which appeared to be working primarily in patients with the relapsing-remitting form of the disease. Furthermore, the interferon trials that didn't work as well may have been because the proper dose wasn't given, and beta interferon could be given in a higher dose than alpha interferon. Marcus ran the idea by his immediate supervisor, who was not impressed by the earlier interferon trials and strongly opposed the idea. Undeterred, Marcus went to Dick Love, the president of Triton Biosciences, to argue his case. Marcus' supervisor was furious that Marcus had gone over his head and threw a temper tantrum. In the end, Love sided with Marcus, the supervisor left the company, and Marcus was given the okay to proceed. "You're in charge," Love said to him.

Marcus called Bob Fishman, then the head of neurology at the University of California, San Francisco, who recommended he visit Ken Johnson (who had now moved to Baltimore). Panitch was with Johnson's group and in the midst of completing his interferon gamma trial. Marcus also spoke to Dale McFarlin at NIH, who provided encouragement. The question was how to proceed, and as it turned out, the meeting to discuss the next step was also held in northern California, this time in the Oakland Airport Hilton, across the bay from the San Francisco airport where Thomas Merrigan had organized the first meeting that led to a trial of alpha interferon in MS.

One of the central questions was whether to do a large phase three trial involving hundreds of patients or to first do a phase two pilot trial. Hill Panitch wanted a phase two trial, arguing that a pilot was needed to establish dosage. Furthermore, there were some properties of the interferon that were similar to gamma interferon, and he was worried that the drug might inadvertently stimulate the immune system and cause attacks. The decision to do a phase two trial turned out to be crucial, as it provided information required for choosing the right dose in the phase three trial. In human drug development, phase one trials are done to test safety, phase two trials to establish dosage and to demonstrate a positive clinical effect, and (if phase two trials are positive) phase three trials are carried out in conjunction with the FDA to obtain approval to market the drug.

The pilot trial was carried out at three centers and involved thirty patients with relapsing-remitting MS. There were six patients per group, and twenty-four patients injected themselves three times a week with 0.8, 4, 8, or 16 units of the interferon. One group was given a placebo. The results? Positive. In the first six months there was a dose-dependent reduction of attacks in the interferon-treated group as compared to placebo. The highest dose resulted in unacceptable side effects (fever, chills, headache, and fatigue) and was abandoned even though patients did best on it. After six months, all patients, except those on the placebo, were switched to the 8-unit dose, and after three years ten of the interferon-treated patients (42 percent) were without attacks, while only one patient in the placebo group (17 percent) was attack-free. To Hill Panitch's relief, unlike the gamma interferon, no exacerbations were precipitated by the recombinant beta interferon. On the basis of the pilot trial, it was time to design the pivotal phase three trial that would determine whether interferon beta-1b (trade name Betaseron) would become the first FDA-approved drug for MS.

When I was learning to design experiments during my first fellowship in Colorado, Henry Claman told me, "If you had an infinite amount of money and an infinite amount of time, you could do an infinite number of experiments and test an infinite number of variables. But that's not how science works, Howard. You have to make choices. You must make an educated guess about which variables to test and what to measure." The advice was in some way analogous to the Bill Rodgers–Howard Weiner marathon experiment that I used to teach my fellows. How to design the experiment to demonstrate a significant biologic difference? The challenge confronting Steve Marcus and his advisors as they sat down to plan the phase three trial of Betaseron was which variables to test—what type of patients to treat, which dose to give, and which outcome measures to establish. These decisions could make or break the trial just as deciding which question to ask to distinguish between myself and Bill Rodgers would determine whether the experiment succeeded or failed.

Which patients to treat? Patients with the progressive form of MS or those with the relapsing form? It was decided to treat relapsing patients. What would be the primary end point? Number of relapses or disease progression as measured by disability? It would be number of attacks. The primary end point was crucial. Approval or disapproval of the drug would be based solely on whether the primary end-point was reached irrespective of the other findings. It was up to the investigators to decide on the primary end point. For approval of a drug, one couldn't measure a series of outcomes and choose afterward which one to use. The primary outcome had to be identified in advance. The number of attacks in patients with the relapsing-remitting form was chosen, as it was felt it would be easier to measure attacks than it would be to assess changes in disability in patients with the progressive form of MS, and progressive patients had fewer attacks.

The next decision was how to give the medication—intravenously, into the nervous system by spinal tap or by a tube connected to a device placed under the scalp, or subcutaneously with the patients injecting themselves? No one felt it had to be given into the nervous system, even though Jacobs had injected interferon into the nervous system via a spinal tap, as it was believed that MS was a disease of the immune system, not just of the brain. The choice was then subcutaneously or intravenously. A debate ensued. Ernie Borden, one of the consultants, argued that it should be given IV, like the gamma interferon in Hill Panitch's trial, as one would get higher levels of the drug into the bloodstream. But the pilot trial showed positive results with the medication given subcutaneously. The decision? Subcutaneously. That way patients could give it to themselves.

But of all the decisions made, the most impactful was the decision to include MRI imaging of the brain in the patients. Don Paty, a neurologist from Vancouver and a pioneer in application of MRI imaging for the study of MS, argued that it was crucial that MRI be performed, to include an objective measure of the disease. Paty had first done MRIs as part of a trial of cyclosporine that Jerry Wolinsky of Texas organized. Wolinsky thought MRIs would be important in MS and asked Paty to do them, as at that time Wolinsky didn't have an MRI machine in Houston. Paty remembers a crucial meeting in Toronto in the midst of a giant snowstorm where he presented his case for MRIs to Steve Marcus. In addition to yearly MRI scans on all the patients, Paty asked to do scans every six weeks on the patients in Vancouver—a total of 881 cranial MRI scans on fifty-two patients. It would add another million dollars to the study. Marcus was an oncologist and was accustomed to studies in which tumors got bigger or smaller depending on the treatment. To him it made sense to see if the areas of the brain involved by MS changed. Thus Marcus approved Paty's request. All he had to do now was to meet with the president of Shell and get approval for the final budget.

The meeting was held on the forty-sixth floor of the Shell Plaza building in Houston. Marcus remembers going through many doors and then being led to a special elevator that brought him directly to the president, John Bookout. "He sat impassively and listened as I presented my case," Marcus said. "Bookout had a mind like a steel trap and only asked occasional questions, but they all went to the heart of the matter. I told him I needed $8,000 per patient, which came to $5.6 million, plus a million for Paty for his MRIs." At that time, Shell Oil was the second largest corporation in America. What was a few million dollars? Bookout liked the idea and after having Marcus repeat his presentation for Dr. Antonio Gotto, a cardiologist then at Methodist Hospital, Bookout approved the trial when Gotto called and said simply, "Sounds reasonable to me." Thus the approval to proceed with the trial that led to the first FDA-approved drug for MS was given by an oil executive and a cardiologist.

Marcus then met with the FDA, which wanted more than one dose used and agreed that one trial would be adequate for approval if the primary outcome was reached. As it turned out, some centers couldn't participate because they didn't have an appropriate MRI facility. Marcus also did something that had a major impact on drug development for MS. He obtained classification of the drug as an orphan drug, since at the time there were fewer than two hundred thousand patients diagnosed with relapsing-remitting MS in the United States. Orphan drug status provided certain tax credits and the possibility of accelerated approval of the drug. More important, orphan drug status helped other companies trying to develop MS drugs, as it granted seven years of exclusivity to market the drug.

However, as the trial progressed, Betaseron itself became somewhat of an orphan, caught up in the midst of corporate buyouts and changes of direction of companies. Cetus, which made Betaseron, was sold to Chiron, and Chiron felt that Betaseron had no commercial value. Then in 1990, Shell Oil decided that pharmaceuticals shouldn't be part of an oil company after all and sold Triton, its pharmaceutical division, to Berlex, a U.S. division of the Berlin-based Schering AG. Berlex bought Triton in order to get rights to an anticancer drug called fludarabine; Betaseron was just a wild card that was acquired with fludarabine. As it turned out, when Betaseron was finally presented to the FDA in 1993 Marcus was not present because he had taken a new job after the sale of Triton. Nonetheless, the trial began. Early in the course of the trial, however, there seemed to be a large number of exacerbations observed. Could the worsening be caused by the Betaseron? A data safety monitoring board was convened and led by Stan van den Noort, a neurologist with expertise in MS who had participated in writing *Therapeutic Claims* and had looked at the data in an unblinded fashion and saw that most of the exacerbations were in the placebo group. The board was not allowed to tell the investigators this, as it would effectively render the entire trial unblinded. The board told them simply, "No need to stop the trial."

At the same time that the trial of Betaseron was being put together, Larry Jacobs was planning how to proceed after obtaining positive results of his double-blind trial in which he had injected beta interferon into the nervous system by spinal tap (intrathecally). The next stage was to carry out a larger trial, and Jacobs prepared a grant proposal that was presented to an NIH scientific committee in a hotel near NIH in Bethesda, Maryland. Jacobs knew it was time to move from testing a naturally occurring interferon that had to be extracted from human cells to a recombinant interferon and he had tracked down an interferon called Bioseron (interferon beta-1a) made by Rentschler in Europe. It differed from Betaseron (interferon beta-1b) by a few amino acids, and was made in different cells. Jacobs proposed a trial that would have four treatment groups: intrathecal natural interferon, intrathecal recombinant interferon, systemic recombinant interferon, and placebo. Cost of the trial: $1 million. Ironically, in the same hotel, a meeting was being held to discuss Betaseron, the interferon Marcus would be testing. Jacobs' proposal was severely criticized by the committee. The committee agreed that interferon should be studied in MS but that the trial was too complex and it was asking the wrong question, testing the wrong variable. The committee felt there was no future for natural interferons in the treatment of MS and that relapses were not an appropriate primary outcome to measure. Jacobs was upset, as he preferred to measure relapses, but he agreed to these points and came back with a $2 million budget that measured disability and incorporated MRI. The review committee liked the revised proposal better but did not want a group treated with intrathecal interferon; they felt that injecting interferon into the spinal canal made no sense. Finally, Jacobs came back with a simple protocol that had only two groups—systemic intramuscular interferon versus placebo. The budget was now $5 million.

The grant was finally approved in 1989 (by which time Steve Marcus had already received approval from Shell and had started his interferon trial—industry had moved faster than government). As it turned out, Bioseron was owned half by Biogen and half by Rentschler, a daughter company. Jacobs went to Biogen, the Cambridge, Massachusetts, company that had approached Hill Panitch to test their recombinant gamma interferon (Immuneron) in MS, the interferon that triggered attacks in MS patients. Jacobs spoke to Jim Vincent, the CEO of Biogen, but there was no initial interest, they were interested in developing a different drug. Finally Jim Vincent agreed to support the trial if it could be done so that it was registered with the FDA. It also helped that the NIH was providing financial support. The Biogen trial finally began in 1990. Everyone thought it was a long shot, especially with disability as the primary endpoint.

ON DECEMBER 21, 1992, Don Paty in Vancouver received a call from Joy Wallenberg at Berlex in Richmond, California. "You need to come down right away," she said. "We've broken the code." Paty's wife wasn't happy that he would be flying to California on December 23, just before Christmas, but he made the trip nonetheless. "The results blew my mind," Paty told me. "There was a clear decrease in attack rate, but what really excited me were the dramatic changes on MRI." With the final results in, the FDA decided to fast-track the approval and scheduled an advisory committee meeting to be held on March 19, 1993. The scientific articles reporting the final results were published in *Neurology,* a major clinical neurology journal in the United States. Most scientific articles undergo two to six months for the review process and revisions and then are published four to six months later. The two articles reporting the clinical and MRI effects of Betaseron were received by *Neurology* on February 6, 1993, and accepted for publication in final form six days later on February 12. The articles were published in the April 1993 issue. Steve Marcus was thanked in the acknowledgment section of the article.

On March 19, 1993, Betaseron was presented to an FDA advisory committee in Rockville, Maryland. The FDA has very specific requirements regarding drug approval: the drug must show positive clinical effects in controlled trials, the company must have reproducible methods to manufacture the drug, and the drug must have an acceptable safety profile. Not all drugs are presented to a scientific advisory committee, only those where there may be controversy, where the FDA needs help, where the drug is the first of its type, or where the FDA wants an independent recommendation before making their decision. The recommendations of the advisory committee are not binding, but the FDA rarely goes against their advice. There are advisory committees for different disciplines, and Betaseron was brought before the peripheral and central nervous system drugs advisory committee.

The room where the advisory board meets is arranged in a specific fashion, almost like a courtroom or a Senate hearing. In the front there is a rectangular table where the advisory committee sits, facing a podium and screen where presentations are made. On one side of the table the representatives of the company sponsoring the drug sit, and facing them sit representatives of the FDA staff. In the other half of the room are advocates for the treatment being considered, the press, representatives of competing drug companies, and Wall Street analysts ready to call in buy or sell recommendations depending on how the advisory committee votes. Television cameras record the hearings, and nowadays one can watch them on the Internet.

The FDA hearing was chaired by Dr. Stanley Fahn, a neurologist with expertise in movement disorders, from the College of Physicians and Surgeons, Columbia University, New York. He introduced the neurologists on the committee. Then the executive secretary of the committee from the FDA began by reporting that all the committee members had filed conflict-of-interest disclosures and were felt not to have significant conflicts. The FDA has a Center for Drug Evaluation and Research (called CDER) and a Center for Biologics Evaluation and Research (called CBER). The division of biologics reviews vaccines and recombinant products, and the division of drugs reviews classic drug molecules. Because Betaseron was made by recombinant technology, it fell under the purview of CBER. As it turned out, Betaseron was the first drug to be reviewed by an intercenter agreement between CBER and CDER, but rules for approval fell under CBER guidelines. This was lucky, because under CBER guidelines, only one independent clinical study was sufficient for approval, whereas at CDER, two were needed. This made the approval of Betaseron easier.

The FDA hearings are open public hearings at which anyone may speak. Ironically, the first person to speak was Larry Jacobs. In retrospect, his presentation was the beginning of what has subsequently been termed the "interferon wars." By this time Jacobs' Biogen-sponsored trial of recombinant interferon beta-1a (now known under the trade name Avonex) was under way. Jacobs told the committee that the primary outcome measured in his trial was disability, not attack rate, because disability was more important, and one should treat the patient, not the MRI scan. His point was an important but controversial one. What is the best way to measure the effect of a treatment on MS? Attack rate? Disability? Changes on MRI scan? All three are linked, but not in a one-to-one fashion—another tentacle of the monster that made MS so difficult to treat. As it would turn out, however, it was the MRI scan that ultimately convinced scientists and physicians that drugs were having a real effect on MS.

The next person to speak from the audience was a woman named Pat Redondo, who suffered from multiple sclerosis and represented the National Multiple Sclerosis Society. She told how every hour an adult was diagnosed with having MS. She herself had been diagnosed twenty years earlier and was now in the progressive phase. "If this drug was available twenty years ago," she said, "then perhaps I wouldn't be progressive. It is too late for me, but not for others."

At this point in the proceedings there were five presentations by members of the FDA followed by five presentations by representatives of Chiron and Berlex, the commercial sponsors. First the mechanism of action of interferon was reviewed: it stopped cells from dividing, it had antiviral effects, and it could affect immune function. There was no clear indication why it was helping MS patients. Then the conditions for which interferons were approved by the FDA were outlined: warts, a type of leukemia called hairy cell leukemia, and chronic hepatitis. It had not been tried in an autoimmune disease such as multiple sclerosis. The manufacturing of the drug was then reviewed. No problems.

Now it was time to review the results, which would determine whether the advisory committee would recommend approving interferon for the treatment of MS. Dr. Janeth Rouzer-Kammeyer of the FDA presented the FDA's review of the efficacy data. "In order to qualify for the study," she said, "the patient had to have at least two exacerbations in the two years preceding the study. The primary end points were defined as the proportion of patients that had no exacerbations (were exacerbation-free) and the frequency of exacerbations. The protocol did not specify whether one or both of these outcomes were necessary to declare a win."

She used a sports analogy: *to declare a win.* Even in the eyes of the FDA, on one level evaluation of the trial was like a game. It would be a win for the MS patients, who would have a new drug to help with the disease. It would be a win for the pharmaceutical company that could market the drug and make hundreds of millions of dollars in profits.

There were three groups in the trial: a high-dose group, a low-dose group, and a placebo group. The annual exacerbation rates were 1.27 attacks per year for the placebo group, 1.17 for the low-dose group, and 0.84 for the high-dose group. The high-dose group saw a 34 percent reduction compared to the placebo group, which gave a p value of 0.0001—only a 1 in 10,000 possibility that it could happen by chance. The other primary outcome being measured, the percentage of exacerbation-free patients, was also significantly different between the high-dose group and placebo. Thus both primary outcomes were positive. However, Betaseron did not affect disability as measured by two disability scales, the EDSS and the Scripps scale.

Next came a detailed discussion of the exacerbations. As it turned out, not all of the exacerbations were verified; almost 20 percent were not. There were also issues regarding patients who had dropped out of the study and the degree to which patients were truly blinded, since those taking the high or low dose had flulike symptoms that weren't present in the placebo group. Dr. Jawahar Tiwari, representing the FDA, summarized his analysis. "We have the two ends of the spectrum," he said. "We have a most liberal interpretation which includes all the patients, includes all exacerbations, having a highly significant difference." He paused. "The most conservative interpretation includes only verified exacerbations and excludes early dropouts as a success. In this case, we have a p value of 0.057." A p value of 0.057 was close, but it was not significant. (According to the funeral joke, it had not become funny.) It would be up to the advisory committee to decide how to best view the data when the final vote for approval or disapproval was made.

Next came a detailed discussion of the drug's toxicity and whether patients and doctors were truly blinded to the treatment. It became clear that the patients were not completely blinded, since only those injecting the interferon had reactions at the site of injection and flulike symptoms. As it turned out, since two doses were tested and there was a difference between the two doses in efficacy, one of the doses served as a control for the other. There were two physicians involved in looking after the patients, a treating physician and a blinded examining physician who didn't know the treatment and didn't speak to the patients. Blinding was crucial. If the committee felt that the blinding was not adequate, it could invalidate the results. But blinding had been done in a rigorous fashion; women patients were even asked to wear slacks for the examination so the examining physician could not tell if there were injection site reactions in the thighs, where the interferon was given. One of the neurologists on the panel wanted to be sure that this was truly the practice. "You would fail your boards in neurology if you didn't take the patient's clothes off," he said.

Dr. Fahn called for a fifteen-minute break, after which Don Paty presented his MRI data, which left little doubt in the minds of the committee that Betaseron had a real biological effect in multiple sclerosis. The linkage of MRI to the disease process in MS is point fourteen of the twenty-one points.

"My name is Dr. Donald Paty," he began. "I am a professor of neurology at the University of British Columbia. We are very pleased to present the MRI data to you. This is a new technology for the evaluation of end points in a clinical trial of multiple sclerosis. However, we have been working on this methodology for about ten years."

Paty began with the results of the fifty-two patients at his institution who were scanned every six weeks, part of the million-dollar MRI experiment that Steve Marcus had gotten Shell to support. The MRI was considered active if there were new, recurrent, or enlarging areas of disease, areas that appear as white spots in the brain. Thirty-five percent of the scans in the placebo group showed disease activity by one or more measures, as compared to 15 percent in the high-dose group and 17 percent in the low-dose group. The active lesion rate was 3 per year in the placebo group and only 0.5 per year in the high-dose treatment group. Although not used by Paty in the Betaseron trial, it is now common practice to inject an element called gadolinium into the vein of patients as a marker when they undergo MRI. If there is active inflammation in the brain, gadolinium leaks into the brain and lights up the inflamed area (this is called gadolinium-enhancing lesions).

Paty then presented results from the entire group of patients in the trial in which burden of disease was measured. "This is analogous to the cancer approach of burden of disease," he said. "Burden of disease" measured the entire area of the brain that was affected by MS and was measured before patients entered the study and after treatment for two years. The burden of disease increased by 20 percent in the placebo group and decreased by 4.2 percent in the high-dose treatment group. Burden of disease was important. In a group of MS patients in whom MRI was performed and then patients followed over a 14-year period, David Miller and his group at the Institute of Neurology in London, England, showed that the larger the burden of disease a patient had when first scanned, the greater the disability fourteen years later.

"Were these MRI analyses done in a blinded fashion," Dr. Fahn asked, "or did the radiologist know the treatment arm?"

"The entire MRI analysis was separate from the clinical aspect and was blinded as to treatment," Paty said. Then Paty described one of the main tentacles of the monster, disease activity was going on in the brain that the patient didn't feel. His group discovered in patients that underwent every four-week MRI scans as part of a natural history study that only about one in five of the new spots that appeared on the scan could be felt in any way by the patient. "The rate of positive scans was about five times the clinical attack rate," he told the committee. It is like having high blood pressure and not knowing about it until it is too late. However, the MRI was not a perfect correlate of disease activity because both doses of interferon reduced the burden of disease visible on the MRI to an equal extent, whereas the higher interferon dose had a more pronounced effect on the relapse rate.

"The MRI data support the clinical data in terms of reduction of disease activity as measured by relapse rate," he said in his summary statement. "More important," he added, "it had an effect on the total burden of disease and therefore really changed the natural history of the disease."

It was time for Dr. Jay Siegel to present the FDA's analysis of the MRI data. Siegel pointed out again that the MRI was a secondary measure of outcome. The primary outcome measured by the trial was attack rate. After detailed discussions and repeat analysis of the data, it was clear to the FDA that there were distinct beneficial changes on the MRI in those patients treated with Betaseron.

There was one curious finding in the analysis of burden of disease: a decrease in the third year of the study in the volume of the brain affected by the MS. Paty smiled sheepishly and explained, "The lesion load was measured by a single technician who traced the areas of MS on the scan. When the numbers she was getting suddenly changed, we began to investigate. She was very consistent in the first two years, and then something happened in the third year. It turned out that in the third year she got bifocals. So we had to have her go back and redo the entire analysis." Later he told me, "The results didn't change. It just cost us time and money."

Henry McFarland, chief of the neuroimmunology branch at the NIH, was asked by the FDA to serve as a special consultant to the advisory panel because of his expertise in MRI imaging in MS. He asked a number of questions regarding the MRI imaging, including positioning the patient's head in the scanner.

Dr. Jeffrey Latts from Chiron presented data on outcome variables, data consistency, and reproducibility and documentation of exacerbations. Latts emphasized that there were many clinically positive findings in the study. Patients in the high-dose group had fewer exacerbations, less severe exacerbations, fewer hospitalizations, less use of steroids for attacks, and a longer interval from the start of treatment to when they had their first exacerbation.

Don Paty spoke again. "I want to make a point here that MS is a bad disease," he said. "We are not dealing with a trivial disease, even though there are benign patients." There was one sticking point, however. If MRI represented the disease process and Betaseron had such a dramatic effect on MRI, why didn't Betaseron affect the disability of MS patients as measured on the disability scales? "I wish I knew the answer to that," Paty said. "I think it is primarily related to the location of the lesions."

Latts finished his presentation for the company by expertly fielding questions from the panel. Why was there a difference in the placebo groups in the United States and those in Canada? Should the trial be considered one trial or two? What primary outcome would they choose to measure if they started over again? Did milder exacerbations occur at certain centers? Did they ask patients for a quality-of-life assessment—

that is, whether the treatment was worth it? How long did the flulike symptoms last? Could the patients have gotten a fever from the shots that led to what was considered an exacerbation? If there was not a significant p value regarding disability, was it worthwhile to have a drug that only had an effect on attacks?

The committee deliberated. It was clear that Betaseron was not a cure; most patients on therapy who were followed long enough eventually had an exacerbation. It was also clear that in the end the presence of positive MRI data convinced the committee that Betaseron was really doing something for MS. Henry McFarland said that it was one of the few trials where everything went in the right direction. He felt over time Betaseron would affect the course of the disease and as MRI imaging became more sophisticated it would be accepted as a primary measure of the disease process. McFarland's statement had a major impact on the panel. Dr. Dennis Choi liked the impartiality of the MRI data because he was worried about patient blinding. Dr. Ira Shoulson echoed the thought, as did Dr. Peter Whitehouse, Dr. Gilman, and Dr. Antonio Delgado-Escueta. The MRI data gave them confidence that Betaseron was having a true biological effect.

When it was time for the vote, Dr. Gilman read the question. "Has the sponsor provided sufficient evidence in patients with mild to moderate relapsing-remitting multiple sclerosis to support the conclusion that Betaseron is effective in decreasing exacerbations in patients with multiple sclerosis?" The vote was seven in favor, two against.

Although the recommendations of the advisory committee are not binding on the FDA, it rarely ignores the advice of an advisory committee, and on that day a new era in the therapy of multiple sclerosis began: an FDA-approved drug with a pharmaceutical company to market it and hundreds of millions of dollars to be made.

The proceedings adjourned at 4:05 p.m., immediately followed by a press conference. The medication became available for doctors to prescribe for their MS patients four months later, in June 1993.

WITH THE APPROVAL of Betaseron, doctors and MS patients were confronted with new choices. Who should go on the medicine? What about progressive patients? The drug was expensive ($10,000 per year), and insurance companies scrambled to establish a mechanism by which to approve payment. If the drug had been shown to decrease exacerbations in relapsing-remitting disease, could it be prescribed for someone with progressive disease? There was another major complicating factor: Berlex was caught by surprise and didn't have enough Betaseron for all the eligible patients. A lottery system was established.

The approval of Betaseron created extra work for the doctors and nurses taking care of MS patients. We had to explain to the patients what the medication was and wasn't. We had to explain how to give the injections and how to handle the side effects of flulike symptoms. For a minority of patients, a troubling side effect of Betaseron was depression. The drug created false hopes and disappointment for those with severe disease. Then there was the emotional trauma of self-injecting a medication every other day with no immediate benefit, only the knowledge that it would help one's disease but not cure it. Nonetheless, the approval of Betaseron was a breakthrough, and as time went on, it was clear that some patients benefited substantially from the drug.

The approval of Betaseron created a crisis for the ongoing trial of the interferon being tested by Larry Jacobs and Biogen. At the time Betaseron was approved, the Biogen trial was in progress, and 301 patients had been enrolled. Half were receiving a weekly intramuscular injection of interferon beta-1a, and the other half were receiving a placebo. Was it ethical to continue with the trial if there was already an interferon proven to help MS patients? What if people secretly took Betaseron? The investigators held an emergency meeting at NIH to decide how to proceed. Steve Hauser was on the advisory committee, and Carl Leventhal represented the NIH. They reviewed the data in detail. Larry Jacobs wanted to continue entering patients in the trial, but Dr. Diane Cookfair, the statistician who headed the statistical center at Buffalo General Hospital, said there were enough subjects in their trial. This recommendation was presented to the safety monitoring committee and approved in July 1993. It was decided to end the Avonex study one year early, and no new patients were entered. The decision was made without knowledge of interim analysis of the drug's efficacy and was based only on the statistician's assessment that a conclusive answer regarding efficacy could be obtained with the data gathered so far. Thus the study was stopped without anyone knowing whether the drug worked or not.

As it turned out, the drug did work, and on December 4, 1995, Avonex was presented to an FDA advisory committee for the treatment of relapsing forms of multiple sclerosis. Stanley Fahn of Columbia was again the chairman of the advisory committee, and, as before, the FDA hearing began with an open public hearing where anyone could speak. At the Betaseron hearing, Larry Jacobs spoke in the open public hearing; now he sat to the right of the head table with the people from Biogen. Dr. Wieland Wolf was the first to speak in the open public hearing, and he continued the interferon wars by bringing it to the committee's attention that the interferon used by Jacobs in the Biogen trial had been made in Laupheim, Germany, under a joint venture between Biogen and Rentschler. However, Biogen would be setting up a different manufacturing process for the interferon they would market and would grow it in a different cell line. "I want to express my serious concerns about the potential for differences between this new Biogen material and the interferon beta-1a produced in Laupheim," he said. He then brought up the example of tPA, an anticlotting factor produced by Genentech, where a change in the manufacturing process led to a less potent drug than the original material. "Since the dissolution of our joint venture, there has been no scientific collaboration between our two companies," he said.

The next speaker in the open public hearing was a representative of the National Multiple Sclerosis Society. She had been diagnosed with multiple sclerosis ten years earlier, just prior to her fortieth birthday. Since then she had had attacks and accumulated disability. She asked the panel to approve the drug quickly and gave a toast in Spanish: *"Salud, dinero, y amor, y el tiempo para disfrutarlos*—health, wealth, and love, and the time to enjoy them." Later in the afternoon a second patient spoke. "You folks are all experts in things related to drug study, to statistical evaluation, to observation of patients," he said. "But I think you do not live with it and you do not really understand what we want is something that will give us some hope to make our lives better and more functional."

The next presentation was from the FDA on the comparison of the drug made by Rentschler and the new interferon produced by Biogen—

the point made by Wieland Wolf. There was no difference biologically, and the FDA considered them equivalent. Then a presentation by Biogen. Avonex was interferon beta 1a, whereas Betaseron was interferon beta 1b. Avonex had an amino acid sequence the same as natural human interferon beta and was glycosolated (had sugar molecules attached to it). John Alam from Biogen presented the data and acknowledged both Larry Jacobs, who was the study chairman, and the National Institute of Neurological Disorders and Stroke or the NINDS, as it is known. The NINDS provided grant support to the investigators and also chose and supported the safety monitoring committee.

As requested by the NINDS at their planning meetings, the primary outcome measure was the length of time it took to sustained progression of disability. Secondary outcome measures were attack rate and MRI imaging that was carried out every six months. To be eligible for the study patients had to have had at least two exacerbations in the three years prior to the study and have only mild to moderate disability on the EDSS disability scale (between 1 and 3.5). Patients with chronic progressive MS were specifically excluded. They reported that the study ended early, with 288 patients completing the study when the sample size was recalculated in the spring of 1993. Only later, upon questioning by the FDA, did it become clear that the reason they recalculated the sample size was because of the positive Betaseron trial. However, there was only one number that mattered. Was there a difference between the placebo and treated group time to disability?

The answer. Yes.

The p value was 0.024.

The treatment also decreased the number of attacks and the number of gadolinium-enhancing lesions visible on MRI.

After a recess, the FDA presented its evaluation. They reanalyzed the data, checked for effects of other medications used in the study, and considered the use of video exams as part of the evaluation. They confirmed the 0.024 p value and pointed out that for reasons not clear there was no change in burden of disease as measured by T2 lesion volume on MRI.

The FDA panel noted that of the four sites that entered patients into the study—Buffalo (the lead site), Cleveland, Portland, Oregon, and Washington, D.C.—the one site that did not show a positive effect of the treatment on disease progression was Larry Jacobs' Buffalo site. If his site wasn't included, the p value went from 0.024 to 0.004. Why didn't the Buffalo site show a positive effect? It turned out that the placebo group did not show as much disease progression in Buffalo as the placebo groups at the other sites. One speculation to explain this difference was that the patients entered at the Buffalo site were at an earlier stage of their disease. As more and more MS trials were performed, it had become clear that patient selection was crucial, and the rate of progression of placebo groups could make or break a clinical trial. This, of course, was the reason for including more than one site in a study, to control for nuances in the patients enrolled at one site. People from Biogen tried to soften the lack of effect on disability at Buffalo by pointing out that there were fewer exacerbations at the Buffalo site. However, if across all four sites there was no effect on disability and only an effect on exacerbations, the drug would not have met the expectations for its primary outcome and could not be approved.

"Another difference, of course," Dr. Fahn, the head of the advisory committee said jokingly, "is the cold weather in Buffalo."

The FDA panel moved on to another topic. After more technical discussion, the FDA said it was satisfied that the interferon manufactured by Biogen was the same as that used in the trial. Because the drug was being reviewed by the CBER, only one pivotal trial was needed for approval. If one took out patients whose exams were only done by video, the trial was still significant.

After lunch the vote was taken on whether the data were adequate to demonstrate safety and efficacy of interferon beta-1a in relapsing-remitting MS. No one seemed to question that a significant clinical effect had been shown. The effect was real, although Dr. Ernest Borden pointed out that the effect was modest.

The vote was unanimous: eleven in favor, none opposed.

The suspense was over and now a series of secondary questions was raised by the FDA and voted on by the committee. Should use of the drug be restricted to less severely affected patients, or could it be used in relapsing-remitting patients with an EDSS greater than 3.5, which was the upper limit of the patients in the study? The committee voted for no restriction based on disability. Could it be approved for patients with the progressive form of the disease, either secondary or primary? No, because there was no evidence that the drug helped patients with progressive MS. However, a trial was in progress to test Betaseron in progressive MS.

The committee was then asked to comment on the MRI results and why there was a change in gadolinium-enhancing lesions but no change in burden of disease, as measured by T2 volume. Henry McFarland from the NIH was present and spoke, as he had done at the Betaseron advisory committee meeting. There was no obvious answer—not enough was known yet about the relationship between different MRI measures and clinical response.

Other questions were voted on. Should the labeling indicate only a two-year treatment course? No. Should the sponsor investigate optimal duration of treatment in a phase IV trial after the drug was approved? Yes. Were videotaped examinations acceptable to determine a primary outcome measure such as disability? Yes. Should the definition of an exacerbation be modified for future trials? No.

Prior to adjournment further studies that were in progress or planned were discussed. There was a study to evaluate a higher dose of the medication, 60 μg as opposed to 30 μg. And there was a trial planned to treat patients after their first attack suggestive of MS. Both trials would ultimately be performed and addressed the crucial issue of early treatment and the importance of interferon dosage.

The scientific article on Avonex was submitted to the journal *Annals of Neurology* on October 5, 1995, in revised form on December 11, 1995, accepted for publication on December 15, 1995, and published in expedited form in March 1996. Even though Jacobs' original hypothesis was that interferon needed to be given into the spinal fluid to affect viral replication, no mention of viruses or an antiviral effect was made in the publication. "The mechanisms of the therapeutic benefits of interferon beta-1a in relapsing multiple sclerosis are unknown," the authors wrote, "but undoubtedly involve its immunoregulatory actions." Possibilities include augmentation of suppressor T cells, inhibition of the effects and secretion of gamma interferon, and increase in the secretion of IL-10, an anti-inflammatory Th2-type cytokine. Jacobs may have been wrong regarding his initial theory about how interferons could help MS, but his energy and perseverance were singular factors in the use of interferon beta-1a for the treatment of MS.

In 1997 another preparation of interferon beta-1a became available for the treatment of MS. Its trade name was Rebif, and it was manufactured by Serono, a Geneva-based firm. Rebif was identical to Avonex but was given in a different dosing schedule, 22 μg or 44 μg given subcutaneously three times per week. Because of the Orphan Drug Act and because Rebif was identical to Avonex, it could not be marketed in the United States for seven years unless "clinical superiority" could be demonstrated. Rebif eventually became available in the United States in the spring of 2002, one year before the orphan drug protection for Avonex would have ended, when the FDA ruled that Rebif had performed better than Avonex in a one-year direct comparison. There was, of course, a major debate between companies over whose interferon drug was indeed superior and a continuation of the interferon wars was carried out by the marketing departments of each company—each side had valid points. Actually, when Avonex was approved, Berlex sued the FDA, arguing that approval of Avonex violated the orphan drug protection Berlex was given for Betaseron. Avonex was ultimately let onto the market because its side effect profile was better than Betaseron's.

Nonetheless, three large-scale trials had shown that interferons could help MS. Despite the interferon wars, what was crucial is that a treatment had been shown to help MS. Still, major questions remained. How were the interferons actually working? How effective were they? How big a step were they toward ultimately curing the disease?

When the Betaseron results were published in *Neurology,* Barry Arnason, one of the investigators who had helped design the trial, was asked to write an editorial. He discussed the decrease in frequency of attacks and the dramatic effects apparent on MRI, and noted that an effect on disability had indeed been shown, although less convincingly than he would have wished. Clearly it was best for relapsing-remitting patients, though it needed to be tested in progressive patients.

Arnason also addressed the big questions—whether interferons help MS, and how they work. Interferons by definition inhibit viral replication, and viral infections are associated with MS attacks. Do the interferons attenuate viral infections or affect a virus that is inside the brain? Arnason concluded that there was no evidence that the interferons worked by their antiviral effects, even though that was the primary rationale for testing them in MS, a point Hill Panitch also made at a lecture he gave in our MS center. The interferons appeared to work by affecting the immune system, by decreasing gamma interferon, by increasing regulatory cells, and by slowing the movement of immune cells into the nervous system (point nine of the twenty-one points). Trials of interferons in MS did not stem from studies on the animal model for MS, since the interferons were species-specific and could not easily be tested in the EAE model. (In small studies after interferons were tested in MS, interferons have been shown to affect the EAE model.) Arnason ended his editorial by stating that regardless of how efficacious the interferons might prove in the long run, "the natural history of MS has been altered favorably, substantially and above all safely. This is, I believe, the end of the beginning," he wrote. "Whether it is also the beginning of the end, time alone will tell."

11

COPOLYMER 1

ON SEPTEMBER 19, 1996, less than three years from the time the FDA approved the first interferon drug for MS, an FDA advisory committee recommended the approval of a different type of drug for MS, a compound originally called copolymer-1 or Cop-1 (now glatiramer acetate) and marketed under the name Copaxone.

There have been two major theories on the cause of MS, the viral theory and the autoimmune theory, and investigations along each of these lines ultimately led to FDA approval of a drug for MS. The testing of interferons was based on the viral theory of the disease. It was because of the possibility that MS was caused by a virus or that viral immunity was abnormal in MS patients that the interferons were tested, even though it turned out that the interferons were effective not because of their antiviral properties but because of their effects on the immune system.

The second major theory, the autoimmune theory, was grounded in EAE, the animal model of MS, and it was because of experiments in the EAE animal model that copolymer-1 was discovered. From the time Rivers first described EAE in 1933, there have been countless articles written about EAE and endless debate on the degree to which EAE is truly a good model for MS. Is MS two letters or three? If one searches the library of medicine in Washington, D.C., there are over 5,000 articles on EAE. Indeed, without EAE, the creation of copolymer 1, its being tested in MS patients, and then approved by an FDA panel would never have happened. Thus the EAE model has proved, despite its faults, to be a crucial tool to understand basic mechanisms of autoimmunity and immune responses against the brain, and a major avenue to discover drugs for the treatment of MS.

The story of copolymer-1 began in 1967 at the Weizmann Institute in Israel. Its discovery and testing involved three scientists, Michael Sela, Ruth Arnon, and Dvora Teitelbaum, all Ph.D.'s who began their work knowing very little about multiple sclerosis. Sela and Arnon had trained in chemistry in the laboratory of Ephraim Katchalski and were experts in synthesizing random copolymers of amino acids. Teitelbaum was a Ph.D. student in the lab. Sela, the senior member of the group, had already published over one hundred articles on the structure of proteins and how they interacted with the immune system when he read about the EAE model and how one could induce paralysis in animals by injecting them with myelin basic protein mixed with an oily adjuvant. Since MBP is a highly charged molecule, his question was whether a synthetic highly charged molecule that resembled MBP in its amino acid composition and that was coupled to a lipid (fatty compound) could mimic MBP and induce EAE in animals. He postulated that the immune response that caused EAE was due to an acidic lipid and that the MBP was only a carrier. It was sophisticated chemistry carried out in a test tube, but the results in animals would be easy to measure. If the experiment succeeded, injection of the copolymer would cause paralysis in the animals just like injection of MBP. A series of copolymers were synthesized and given numbers—Cop-1, Cop-2, Cop-3. For a year Teitelbaum injected the copolymers into animals. The results? Failure. None of the animals got sick.

At this time the team became aware of the work of Elizabeth Roboz Einstein, a neurochemist who worked in the EAE model (and, incidentally, was Albert Einstein's daughter-in-law). She was among the first to show that MBP or modifications of MBP could be used to treat animals given EAE. The team thought maybe their failed experiment could be salvaged by using the copolymer for a different purpose. Teitelbaum spoke to Sela and Arnon, and they proceeded to test the copolymers for the opposite effect for which they were synthesized. If the synthetic molecule they constructed could not induce EAE, could it suppress EAE? The answer? Yes. It actually worked in the very first experiment. They had started out by trying to design a molecule that could cause EAE and ended up with one that suppressed EAE.

What to do next? They observed one of the cardinal rules of science: repeat the experiment. Thus they quickly made another batch of Cop-1 and tested it. They found that it worked just as well a second time. It was immediately apparent to the group that Cop-1 had the possibility of one day becoming a therapy for MS, and the next experiments were done with an eye to testing it in people. First, did injection of Cop-1 cause a general suppression of the immune system? If it did, it might have untoward toxic effects. The answer: no. Second, if Cop-1 was to ultimately work in people, it needed to suppress EAE in other animals besides the guinea pig. They tested it in mice, rabbits, monkeys, and baboons. To their joy, it suppressed EAE in all species tested.

What exactly is Cop-1 and how does it work? Cop-1 is not that easy to explain, even to scientists, because it is not a conventional molecule. Perhaps the best way to explain Cop-1 is to begin with an explanation of amino acids and MBP. Amino acids are the building blocks of proteins. There are twenty amino acids; like letters of the alphabet, they are linked together to form proteins. Proteins are longer than words, however—myelin basic protein is 170 amino acids long (which is an average-sized protein), for example—and the amino acids of each protein are linked together in a specific sequence, with some amino acids being used more than once. Cop-1 is made up of only four different amino acids, alanine, lysine, glutamic acid, and tyrosine, which are linked together in a random order but in a specific ratio. Thus in Cop-1, for every six alanines, there are approximately five lysines, two glutamic acids, and 1 tyrosine (the exact ratio is 6.0–4.7–1.9–1.0). Thus Cop-1 is not a single protein but different proteins that contain the same four amino acids randomly mixed in a specific ratio. What's most important is that Cop-1 suppresses EAE. Some scientists have difficulty understanding or studying Cop-1 because, as a random mixture, it does not have only one structure. This also created problems for manufacturing the drug because it was difficult to prove that the identical mixture was being made every time. Indeed, for many years every batch had to be tested in the EAE model before being given to people.

In 1971, when the first article was published showing that Cop-1 suppressed EAE, our basic understanding of immunology was in its infancy. Nothing was known about how T cells were triggered to cause EAE, the chemical factors (cytokines) released by T cells hadn't been identified, and all the proteins that caused EAE were not known. Indeed, the scientific understanding of how Cop-1 works has paralleled our understanding of how the immune system works. Although it was not known how Cop-1 mimicked or was cross-reactive with MBP, it consistently suppressed EAE. Thus, over the past thirty years the story of Cop-1 has proceeded along two tracks: its clinical development for the treatment of MS patients, and scientific investigations to understand how it affects the immune system.

The clinical development of Cop-1 was not dependent on understanding its mechanism of action. The logic that drove the clinical development was simple: Cop-1 worked in the animal model of MS; Cop-1 somehow affected the immune system; Cop-1 was not toxic; there were no treatments for MS; test Cop-1 in MS.

The first clinical trial of Cop-1 was conducted in Israel at the Hebrew University Hadassah Medical School by Oded Abramsky, a neurologist who also received his Ph.D. under Arnon at Weizmann and who now chairs the Department of Neurology at Hadassah. Four MS patients in terminal stages of the disease and three patients with a form of allergic encephalitis were treated. They were given 2–3 mg of Cop-1 two to three times a week for four to six months. These patients were true guinea pigs. No beneficial effects were observed, but most important, there were no side effects. Arnon then searched for other MS clinicians to test Cop-1 in MS patients. She even made a film that showed how baboons given EAE were successfully treated with Cop-1. Arnon was able to interest Helmut Bauer, a German neurologist she met at a scientific meeting in Europe and who had a large clinic for the treatment of MS. Dosing was a big issue and was discussed at a MS scientific meeting held in Gottingen, Germany, in 1980. Bauer was in the midst of an open-label trial in which ten patients with mild disease received a daily dose of 2 mg of Cop-1 for a month, and eleven with more severe disease were given a higher dose of 20 mg. Jonas Salk was at the conference and reported on his trials of injecting MBP into MS patients and how he was trying to build up to a higher dose of MBP. The advantage of Cop-1 over MBP was that injection of Cop-1 into animals didn't cause disease. Arnon reported that they were also increasing the dose of Cop-1. Bauer never formally published his results, and although there may have been a suggestion that some patients with early disease improved with Cop-1 injections, nothing could be concluded from his short open-label trial. Bauer's most important results were that he observed no toxicity when a higher 20 mg dose was given.

The breakthrough clinical trial for Cop-1 was conducted in the United States by Murray Bornstein at the Albert Einstein College of Medicine in New York. Bornstein was a clinical MS specialist and investigated factors from the serum of MS patients that were toxic to myelin in tissue culture. It was because of his scientific work with these antimyelin factors that he became interested in Cop-1. He was a flamboyant individual who championed the testing of Cop-1 in MS—to the point where he injected himself with every batch of Cop-1 that was prepared at the Weizmann Institute to make sure it was safe before giving it to patients. Being a diabetic, for him it was just one more injection. Bornstein's enthusiasm and perseverance were critical in the clinical development of Cop-1.

Bornstein was joined by Aaron Miller, a neurologist who now heads the MS center at Maimonides Medical Center in Brooklyn. Bornstein first conducted a preliminary open-label trial that involved sixteen patients, four with relapsing-remitting disease and twelve more severely affected patients with chronic progressive disease. He published the results in 1982. He began by injecting patients with 5 mg per day of Cop-1 for five days, with the plan to decrease the frequency of injections to once per week, but he ended up increasing the dose to a daily injection of 20 mg based on his clinical impression that patients needed continued treatment. Patients were initially injected into the muscle and later under the skin. Although the results were anecdotal, Bornstein truly believed he had observed a positive clinical effect of Cop-1 in some people. Five of the patients seemed to benefit, and two of the relapsing-remitting patients stopped having attacks. From the standpoint of the clinical development of Cop-1, what was most important is that he had established a dose (20 mg per day); and he was determined to proceed.

Next came the big step of carrying out a double-blind, randomized, placebo-controlled trial to establish whether there indeed was a clinical effect of Cop-1 in MS. Bornstein led two trials, the first in relapsing-remitting patients and the second in progressive patients, both supported by the National Institutes of Health, though he had to resubmit his grant more than once until it was finally funded. Close to one thousand patients were screened to find fifty who were eligible for the relapsing-remitting trial. Twenty-eight batches of Cop-1 were prepared, twelve at the Weizmann Institute and sixteen by Bio-Yeda, a small biotech firm associated with the Weizmann Institute. The batches were all tested for their ability to suppress EAE prior to being released for humans. The dose was 20 mg per day, and patients were treated for two years.

The results of the trial in relapsing-remitting patients were published five years later in the *New England Journal of Medicine* on August 13, 1987. The article was entitled, "A pilot trial of Cop-1 in exacerbating-remitting multiple sclerosis." As we have seen, publishing a positive clinical trial on multiple sclerosis in the *New England Journal of Medicine* generates interest and publicity. The article was felt to be of such importance that an editorial accompanied the article and I was asked to write the editorial.

What did the study show? The trial found that the twenty-five patients with early relapsing-remitting multiple sclerosis treated with Cop-1 for two years had fewer MS attacks (sixteen attacks in the Cop-1 group versus sixty-two attacks in the placebo group), and more patients who were treated with Cop-1 were exacerbation-free than those treated with placebo (56 percent versus 26 percent). The data were very clear—in fact, to some in the field they looked "too good." We all knew that even though a trial is done carefully there can be artifacts that lead to false results. One problem with the Cop-1 trial was that the control patients had many attacks during the period of the study, and such a large number of attacks had a major impact on the statistical analysis. Furthermore, no effect on disability was observed in patients in later stages of the disease, and the question was raised whether the study was truly double-blind, as some patients who injected themselves under the skin with Cop-1 had more redness at their injection site than those who injected a control solution. Nonetheless, there was a statistically significant reduction in MS attacks with minimal toxicity, and the study was carefully conducted. Of course, only fifty patients had been studied, and Bornstein and colleagues ended their report with an appropriate word of caution: "These results suggest that Cop-1 may be beneficial in patients with exacerbating-remitting MS, but we emphasize that the study is a preliminary one and our data require confirmation by more extensive clinical trials."

In my editorial I tried to strike a balance between the excitement over the results and the preliminary nature of the findings. A nontoxic immune modulator that could be given early in the disease would be a major advance. However, assuming the clinical results could be repeated in a larger trial, a central question remained: how did Cop-1 work in MS? Some investigators had not found that Cop-1 was related to MBP in humans, and others reported that Cop-1 triggered lymphocytes from all people, irrespective of whether or not they had MS. These immune findings were different from those reported in animals. I pointed out that the role of MBP in MS and the relationship of EAE to MS were not known. I closed my editorial as follows: "Implicit in all clinical trials currently being carried out in multiple sclerosis is the question whether effective treatment for the disease can be developed without knowledge of its exact cause. The answer may be a tentative yes with clinical trials themselves teaching us about the disease process." Clinical trials are experiments in which one directly confronts the disease, and if a positive result is obtained, it moves us closer to both an effective treatment and a better understanding of the disease. Indeed, the first approved drug for MS, interferon beta-1b, was tested, based on what turned out to be an incorrect rationale, but its positive effects have helped lead the way to a better understanding of MS. Similarly, we have learned about the role of the immune system in MS from studies of chemotherapy-type drugs that have an ameliorating effect on the disease.

What was the next step for Cop-1? At the time the first trial was published, a second trial was in progress in which patients with chronic progressive disease were being studied in a placebo-controlled, double-blind trial, at a daily 30 mg dose. The trial in progressive patients was a two-center study carried out at the Albert Einstein School of Medicine and at Baylor University in Houston. Had the results been positive, it would have conclusively established that Cop-1 was an effective treatment for multiple sclerosis. Unfortunately, the results were disappointing. In a trial of progressive MS, the primary outcome measured is how much a person worsened on the neurological disability scale, not number of attacks. Worsening by 1 or 1.5 units on the neurological disability scale was observed in 17.6 percent of the Cop-1-treated patients and in 25.5 percent of the patients who received placebo. Thus there may have been a hint that the drug was working, but the results were not statistically significant. As it turned out, the trial was positive at the New York center and negative at the Texas center. A major problem with the trial was that there appeared to be a large placebo effect, in that patients in the control group did not worsen as much as was expected.

The negative results of the Cop-1 trial of progressive patients meant one of two things: either the drug was not effective in MS, or it was effective but only in relapsing-remitting patients, not in later stages of the disease. After all, in the Bornstein trial it appeared that people with lower levels of disability did best. Additional trials were later done in Israel on 271 patients to study the effect of Cop-1 in relapsing-remitting patients and to study its immunological effects in people, but these were open-label trials. Thus if Cop-1 was ever to be a treatment for MS, a large, double-blind, placebo-controlled trial in relapsing-remitting multiple sclerosis was needed.

In addition to the millions of dollars required for a large-scale phase three clinical trial to be carried out under FDA guidelines, there had to be a reliable method to manufacture the drug. Preparation of Cop-1 was not easy, and from 1979 to 1989 it was done at the Weizmann Institute by a person described by Ruth Arnon as a "supertechnician," Israel Jacobson. After the positive results of the Bornstein trial, Sela brought his wife, Sarah, and a slide projector to the house of Eli Hurwitz, the CEO of the Israeli pharmaceutical company Teva, for dinner and a presentation. Teva was primarily involved in the manufacture of generic drugs, but that night Hurwitz decided to take on the development of Cop-1, a proprietary drug. "It was crazy that I approached him," Sela told me in one of our conversations, "and it was even more crazy that Hurwitz accepted." But Hurwitz did accept, and it took over two years for Teva to develop a method by which Cop-1 could be produced reliably in large quantities.

The final chapter in the clinical development of Cop-1 was now completely out of the hands of Sela, Arnon, and Teitelbaum, and they watched from the sidelines as a phase three trial was conducted in the United States by Teva, which had bought the rights to develop the drug, which it had named Copaxone. It was a double-blind, placebo-controlled trial carried out in 251 patients. To be eligible for the trial a patient had to have had two or more well-defined MS attacks in the two years prior to randomization. Patients received either a daily injection of 20 mg of Cop-1 or placebo. Ken Johnson, chairman of neurology at the University of Maryland, who had also been involved in clinical trials of beta interferon, coordinated the eleven-center clinical trial in the United States. He had met Arnon and Irit Pinchasi from Teva at an MS scientific meeting in Rome. There was a delay in beginning the trial because the FDA would not allow the EAE model to be a biological marker for release of Cop-1. Ken Johnson wanted MRI to be part of the study, but Teva rejected the idea. First, there were not enough funds; second, they were planning to use the Bornstein trial as one of the pivotal trials for presentation to the FDA, and Bornstein had not performed MRIs. Like the phase three trials for the interferons, the phase three Cop-1 trial was rigorous. There were two neurologists for each patient, a treating neurologist and an examining neurologist; the examining physician didn't know what the patient was receiving. Patients were examined every three months. A nurse coordinator distributed the medication, and relapses were carefully evaluated. An independent data monitoring committee followed the trial for safety. If the trial didn't work, it would be hard if not impossible to argue that Cop-1 helped MS patients.

Entry into the trial began in October 1991, and after 284 patients were screened for eligibility, 251 were randomized to receive either Cop-1 or placebo. Apart from the safety monitoring committee and the statisticians, no one knew the results until a hot summer day in 1994 when the results were unveiled at the Teva facility in Kulpsville, Pennsylvania. Sela and Arnon came from Israel, as did some Teva executives who hated to travel but could not miss being at the announcement. Twenty-five years of work on Cop-1 came down to a single moment.

The results? Positive! There was a 29 percent reduction in the relapse rate in the Cop-1 group compared to a placebo. The p value was significant at 0.007.

Yafit Stark, who managed the trial for Teva, greeted Irit Pinchasi two minutes before the announcement with a kiss and the words "We have a product." Irit came up with the name for the new drug, Copaxone, by combining the word *copolymer* with *axon,* since they felt Cop-1 could protect the axon or nerve fiber from damage by its anti-inflammatory properties. A manuscript describing the findings was submitted for publication to the journal *Neurology* on April 27, 1995, accepted for publication only four days later, and published in July 1995. Now, only one last hurdle remained: the FDA had to approve Copaxone.

The FDA panel met on September 19, 1996, in an open hearing at the Holiday Inn in Gaithersburg, Maryland. The FDA hearing was chaired by Dr. Sid Gilman, chairman of the department of neurology at the University of Michigan Medical Center. I attended the presentation. As is the usual practice, the FDA advisory panel sat at a table in the middle of the room, flanked by representatives from Teva on one side and representatives of the FDA on the other. In the audience, there were representatives of competing drug companies, the press, Wall Street analysts, and advocates from the MS Society. Dr. Gilman introduced the panel and then told them that they had to answer two questions: first, did the two clinical trials that were to be presented by Teva show that copolymer-1 was an effective treatment for MS, and second, was the treatment safe? Paul Leber from the FDA emphasized that not every drug is brought to an advisory committee, only drugs where the FDA feels there is controversy and the answers to the questions about its efficacy and safety are difficult. Leber's words caused people from Teva to shift uncomfortably in their seats.

Dr. Russel Katz from the FDA spoke next. He pointed out that two trials would be evaluated: the Bornstein trial, published in the *New England Journal of Medicine* in 1987, and the Teva-sponsored trial, published the previous summer. He then raised questions about the trials. Teva was not involved in the Bornstein trial but subsequently obtained as much data as possible from the trial and carried out analysis. Katz said that two patients from the placebo arm of the Bornstein trial were not included in the analysis published in the *New England Journal of Medicine* (they were dropped from the study for psychological reasons). "If these two patients were included as part of an 'intent to treat' analysis," he said, "the p value went from 0.038 to 0.18, no longer statistically significant." Another problem related to two documents discovered as part of the document trail, the first of which said the Bornstein trial was designed to be forty patients, not the fifty reported in the *New England Journal of Medicine.* "It is possible," Katz said, "that the sample size was increased on the basis of an interim analysis." This could invalidate the statistics of the Bornstein trial. Katz also pointed out that although the large Teva-sponsored trial was positive in the primary outcome being tested, exacerbation frequency over two years, it was negative in other measures, including the primary measure in the Bornstein trial, proportion of exacerbation-free patients. The results in the smaller Bornstein trial were much more positive than in the larger Teva trial. Finally, he raised a question about chest pain that was experienced in 26 percent of those treated with copolymer-1 but in only 10 percent of the placebo group.

It was time for Teva to present its case. Dr. Carole Ben-Maimon, a senior vice president from Teva, led the discussion. "Copaxone represents a new class of immunomodulators," she began, "that we believe is a unique therapeutic option for patients with MS." She then introduced Ken Johnson, the neurologist who had led the multicenter trial in the United States. He described the different stages of MS and pointed out that the number of attacks a person has at the beginning of the illness is related to later disability. He explained the theory that MS is an autoimmune disease initiated by T cells and that the goal of MS therapy was to limit the number and severity of attacks and prevent disability. He then referred to his fifteen-year involvement with testing interferons in MS in five different controlled trials. "In my view," he closed, "there is a major unmet need for effective therapies, when we have only the interferons at the present time." Although he did not intend it, his presentation initiated the beginning of Teva's marketing strategy for Copaxone versus the interferons.

Ben-Maimon then reviewed the animal studies and early clinical data for Copaxone and how Bornstein had arrived at a dose of 20 mg. She described how Sela, Arnon, and Teitelbaum had first created copolymer-1 to mimic MBP, and hypothesized about possible mechanisms by which it worked, but concluded by saying that "the exact mechanism of Copaxone is not understood."

Dr. Gilman interrupted to ask about reactions that a small number of patients experienced after injection of Copaxone, including chest tightness, anxiety, flushing in the face, and trouble breathing, which lasted about fifteen minutes. The reactions, which were first described by Bornstein and also observed in the Teva trial, and the general safety of the drug were discussed several times during the proceedings. The cause of the reactions was unknown but didn't appear related to the heart or appear life-threatening or dangerous. Two people had died while on Copaxone, but their deaths were felt to be due to unrelated illnesses. There was no evidence that copolymer-1 caused cancer in animals. In the end, Copaxone appeared to be a safe drug with minimal side effects.

Dr. David Drachman asked about the structure of copolymer-1 and whether there was a specific structure of the molecule that was active, since it was a random mixture of amino acids. Ben-Maimon said that after injection under the skin, the copolymer was broken down into smaller pieces at the site of injection. Ruth Arnon added, "We have tried very, very carefully to fractionate the molecule in many, many ways, but we have never been able to come up with any particular fraction or sequence that has a better ability than others."

Ben-Maimon then reviewed the clinical trials and the number of people that had been treated with Copaxone. Aaron Miller was there to answer questions about the study published in the *New England Journal of Medicine,* as Bornstein was deceased. Miller reported that they had screened over nine hundred people for the trial from questionnaires they received after a public call for patients that Bornstein had made in an effort to find patients with very active disease. Then the issue came up of the two patients who had been dropped from the study. The first had psychiatric problems in dealing with having MS and had become nonfunctional by being in the trial, Miller reported. "As far as the second patient," Miller said, "and this is a somewhat embarrassing confession to make, after several months in the trial I became convinced all of her findings were psychogenic. She did not have multiple sclerosis."

Miller pointed out that the study was done at a time when MRIs were not in common use for MS trials.

"Didn't you require hard neurological signs?" Dr. Gilman asked.

"We required neurological signs," Miller replied, "but I am sure you too have been fooled in the past about neurological findings that were not physiological in origin."

"Less as I have gotten older," Gilman replied to laughter from the audience.

The effect of these two patients on the analysis was then discussed in detail. They had been dropped from the study without knowing what treatment group they were in. The advisory committee then reviewed whether forty or fifty people had been originally planned for the trial and the submitting and resubmitting of Bornstein's grant to NIH for funding after the initial submissions were not approved. The panel pointed out that the Bornstein trial was performed as an academic trial, and the *New England Journal of Medicine* would not have published it had the journal not felt it had merit; there were precedents of published studies taken from the literature and presented as pivotal trials for new drug applications. Furthermore, it was uncommon for noncommercial sources to do randomized trials with the rigor and attention to detail of industry.

After an 11 o'clock break, Ben-Maimon presented the Teva-sponsored multicenter trial that had been published the summer before. The entry criteria were at least two relapses in the two years prior to entry—identical to the criteria for the interferon trials. The primary outcome being measured was the mean (average) number of observed attacks during the two-year trial period. Patients who felt they were having an attack had to be examined by a neurologist within seven days of reporting the attack. Those who were first to enter the trial continued on medication or placebo in a double-blind extension trial. The primary end point was met. At twenty-four months the annualized relapse rate for treated patients was 0.65 and 0.84 for placebo, a statistically significant difference. Other analyses also favored Copaxone: time to first relapse and proportion of relapse-free patients. In terms of disability, significantly more patients receiving Copaxone were found to have improved on the disability scale and more receiving placebo worsened (the p value was 0.037). MRI was then discussed. MRI imaging was not performed as part of the trial apart from pilot studies done on fourteen Copaxone-treated patients and twelve placebo-treated patients. Nothing could be concluded from those, and Copaxone would be approved or disapproved by the FDA with no MRI results. Nonetheless, a thirty-center study on the effects of Copaxone as seen on MRI was then carried out in Europe and Canada. The results of this MRI study would be crucial for acceptance of Copaxone by physicians, since the interferons showed positive effects on the MRI. The discussion again turned to issues of safety and side effects in the Teva trial, including local injection site reactions and the flushing reaction and tightness in the chest that occasional patients experienced.

Ben-Maimon summarized the two trials as showing both a reduction in relapse rate and changes in disability, then turned the floor over to Dr. Jerry Wolinsky. Wolinsky headed the MS program at the University of Texas in Houston and had participated in the Copaxone trials. Wolinsky spoke of the enormous cost to patient and society for someone who has MS and how we needed additional drugs to help treat MS. Drachman then asked Wolinsky what he prefaced as perhaps an unfair question. "Given what is now available and given a patient with MS, where would this treatment fit in if it were approved?" With the disclaimer that it was his own personal opinion, Wolinsky said that Copaxone would be a good choice for a recently diagnosed patient as an alternative to interferons and for a patient concerned about side effects that occurred with interferons or who was having significant side effects on the interferons. At the break, Wolinsky told me he had tried to be as neutral as possible in his response, stating the obvious, so that he wouldn't be viewed as too "pro-Teva." Nonetheless, Wolinsky's comments too foreshadowed marketing strategies for Copaxone versus interferons. Then a question that foreshadowed combination therapy in MS (point nineteen of the twenty-one points) was asked by an advisory panel member: "Could interferons and Copaxone be used together?" Ruth Arnon answered that there was no data in animals since human interferons couldn't be given to animals, but it was a theoretical possibility. As it turned out, a large multicenter trial of Copaxone plus Avonex funded by the NIH began in 2004.

The morning session adjourned at 12:53 p.m. and reconvened an hour later. It was a tense session, and in the opinion of many was not particularly favorable to Teva.

After lunch, Dr. Gilman asked Ben-Maimon whether Teva had anything to add.

Answer: no.

Did the FDA have anything to add?

No.

It was time for the advisory committee to debate and either approve or disapprove Copaxone as a new drug for MS. As previously noted, although the decision of the advisory committee is not binding on the FDA, the FDA rarely fails to accept the advisory committee's recommendation. As Gilman opened the discussion, Michael Sela, Ruth Arnon, and Dvora Teitelbaum sat nervously, not sure what to expect. Yafit Stark and Irit Pinchasi dug their hands into those of Aaron Schwartz, a VP from Teva.

Gilman began by saying he felt there indeed were two trials that showed a reduction of frequency of relapses, with the Teva-sponsored trial clearly the better of the two. He was less convinced that an effect on disability was shown, and he was bothered by the fact that changes in disability in Copaxone-treated patients appeared to occur primarily at the beginning of the trial. Dr. Patricia Coyle, the only physician on the panel with expertise in MS, agreed.

Following a discussion of the effect of Copaxone on disability, Dr. Drachman spoke. "We are always hoping that we are going to see a drug that is as good as penicillin was for pneumonia," he said. "It is always disappointing when it turns out that we are dealing with effects that may not be curative." However, he stated there was clear evidence that Cop-1 helped MS patients and it had minimal toxic effects. There was a need for new drugs in MS, and Copaxone appeared safe; he reminded all of the dictum to physicians to do no harm.

Dr. Patricia Coyle, a clinical MS expert from the State University of New York at Stony Brook, added that cutting down relapse rates is beneficial for MS patients and that doing so may decrease disability. Dr. Zaven Khachaturian emphasized that even if only 10 percent of the four hundred thousand MS patients in the United States were helped by Copaxone, it was important, given that there was little on the market for MS patients. "Shouldn't we be looking at it from a wide-angle lens rather than narrowing in on the small?" he asked.

Perhaps one of the most dramatic parts of the afternoon came next: presentations by seven MS patients, sponsored by the National Multiple Sclerosis Society. The patients had been treated with Cop-1 as part of the clinical trials, and they told poignant stories that put a human face on the disease. They gave the names of their spouses and children. They told of not having further attacks while on the medication. They described the flushing reactions and said the reactions weren't bad. They invited the panel to question them. One patient reported she had been injecting herself for four years with no further attacks. Another said that she had had a daughter born while she was taking the drug. A third said she had been on a beta interferon drug but couldn't take the interferon because of side effects; since she had begun taking Cop-1, she felt better. And finally there was a ballet dancer who had had to give up dancing after twenty years of preparation; since taking Copaxone, she was now able to play her violin and go swimming. Michael Sela later told me how touched he was by her story.

Nonetheless, no matter how poignant such stories are, they must always be viewed with caution. Anecdotal stories are told by countless people who take treatments of no proven benefit. The bottom line was simple: had copolymer-1 been shown in appropriately controlled clinical trials to help MS patients? It was on this question that the advisory committee would now vote.

Dr. Gilman asked if the sponsor (Teva), the FDA, or anyone else in the audience wished to speak or make a final statement. Silence.

"Then it is time for the committee to vote on this agent," he said. "To provoke discussion, let me lead off. I feel convinced that the data we have heard today provide two well-controlled, double-blind studies indicating that the medication reduces the frequency of relapses. I have heard information suggesting it is a safe drug. So I will start by saying that I will vote to approve the drug."

The only discussion of substance from the panel was to clarify that Copaxone was being approved for its effect on relapses, not on disability.

With a show of hands, Copaxone was unanimously approved by the advisory committee. The meeting was adjourned at 2:54 p.m. In the end, the afternoon session had lasted only an hour. The mood of the panel had changed in the afternoon and Hill Panitch asked me what the panel had had for lunch. Ruth Arnon and Dvora Teitelbaum had tears in their eyes.

Michael Sela breathed a long sigh of relief. Eli Hurwitz gave Yafit Stark and Irit Pinchasi a hug and a kiss. "I have a policy never to kiss any female employees," he said, "but this is an exception."

A major issue for physicians at the time of approval of Copaxone was that, unlike the interferons, there were no good data available on the effect of Copaxone on MRI measures of disease. These anxiously awaited results were presented at the American Academy of Neurology meetings in Toronto in the spring of 1999 and published in 2001. A double-blind, placebo-controlled trial involving 239 patients showed that Copaxone (now referred to as glatiramer acetate instead of copolymer-1) positively affected MRI measures of disease, decreasing both gadolinium-enhancing lesions and T2 lesion burden.

Although the compound synthesized over thirty years ago, when the odyssey of copolymer-1 began, is identical to the compound now used to treat MS patients, an understanding of how it works has undergone major revision as the latest immune theory has been applied to an understanding of its mechanism of action. The immune theory is complicated but can be summarized as follows. Copolymer-1 binds to specialized molecules on cells whose function is to then trigger T cells. The cells are called antigen presenting cells and the molecules that copolymer-1 binds are called MHC class II molecules. With copolymer-1 bound to its surface, the antigen presenting cells trigger human T cells to become cells with anti-inflammatory properties. The T cells then migrate to the brain, where they suppress the brain inflammation that causes MS by releasing anti-inflammatory substances such as IL-10 and TGF-beta. Copolymer-1 is not simply an MBP analog, as originally designed by Sela and Arnon. Although it may have some cross-reactivity with MBP, it has much broader properties. Human T cells from all individuals can be stimulated by copolymer-1, and after stimulation they shift from Th1 to Th2/Th3-type cells, which act as suppressor or regulatory cells. When they enter the brain, they suppress inflammation via bystander suppression, the basic mechanism we discovered in our work on oral tolerance (point seventeen of the twenty-one points). This is why it can suppress so many types of EAE. Thus, the copolymer designed to cause EAE in mice ultimately worked in MS because of unique interactions with human T cells that weren't known when the molecule was synthesized.

In the end, what has copolymer-1 taught us about MS? How has it moved us closer to the day that we have a cure? Because it is a specific immune modulator, it strongly supports the immune hypothesis of MS and the Th1/Th2/Th3 paradigm. Copolymer-1 validated the EAE model of MS, even though not all drugs that work in EAE help MS, and provides impetus for testing additional drugs in the EAE model.

Nonetheless, since its approval by the FDA, the further clinical development of Copaxone has encountered setbacks. A large, placebo-controlled, double-blind trial in almost eight hundred patients with the primary progressive form of MS was negative. Remember, however, there have been no positive trials in the primary progressive form of MS, which appears to be a subtype of MS that is not impacted by currently approved drugs. In addition, a fifteen-hundred-patient trial of oral Copaxone was negative, most probably because the dosing and oral formulation were not optimal. Further trials of oral Copaxone are being planned; if successful, they will facilitate combination therapy of Copaxone with other drugs. Despite these setbacks, it has been discovered that Copaxone may have additional effects besides immune modulation. In animal models, Michal Schwartz has shown positive effects in preventing nerve damage, including in a model of ALS. Thus one day it could find uses in other diseases besides MS.

In the end copolymer-1 has turned out to be a remarkable mixture of peptides. Like so many discoveries in science, it involved serendipity, careful observation by scientists, and tenacity. The time from its discovery to FDA approval, almost thirty years, is not unusual for the translation of basic science to final clinical application. It has been the first success in our search for immune-specific therapies for MS—the approach that Salk and others have tried and which is continuing to be investigated. Immune-specific therapies are likely to one day be part of the cure. "I'm just a chemist," Sela told me, "but it was so wonderful to hear the ballerina's story." Ironically, the next drug to be approved by the FDA for the treatment of MS was a chemotherapy drug, the opposite end of the spectrum from glatiramer acetate, a global as opposed to a specific immune modulator. Nonetheless, for MS to ultimately be controlled, something stronger was needed when glatiramer acetate or the interferons didn't work.

12

NATALIZUMAB AND BEYOND

A MAJOR CHAPTER in the story of MS was written in 2004 with the FDA approval of a new drug called Tysabri. Tysabri is the trade name for natalizumab, a monoclonal antibody that blocks entry of T cells into the brain. The effectiveness of natalizumab supports point eleven of the twenty-one points, which states that "MS is driven by T cells that continually migrate into the brain and spinal cord from the bloodstream." By blocking T cells from getting into the brain, natalizumab had a profound effect on MS relapses and activity on MRI as measured by gadolinium-enhancing lesions in the brain. Natalizumab presented a new therapeutic approach for treating MS. As it turned out, the story of natalizumab has been a roller-coaster ride for the MS scientific community, for MS patients and for Wall Street.

Where does this story of natalizumab begin? Just as with the development of copolymer-1 (Copaxone) for MS, the development of natalizumab was based on EAE, the animal model for MS. Remember that EAE is caused by T cells that migrate from the bloodstream into the brain where they attack the myelin sheath and cause damage. In the late 1980s, scientists began to identify molecules on the surface of T cells that were required for them to home to specific tissues. One of the basic features of biologic systems, whether it be the biology of T cells or social interactions among people, is recognition and communication. Thus, the creation of natalizumab began with the discovery of unique molecules on the surface of the T-cell surface that allowed T cells to communicate with blood vessels so they could leave the bloodstream and enter the brain.

Just as a key fits into a lock, there are two structures involved in the communication between a T cell and a blood vessel: one is on the T cell and the other is on the blood vessel. The structure on the T cell was identified by a scientist named Martin Hemler in 1987 and is called VLA-4. VLA-4 stands for "very late antigen-4" and, simply translated, is a structure that appears on T cells at a late stage after a T cell is stimulated. When VLA-4 was described in 1987, we had studied it in MS patients and found that VLA-4 was elevated on MS T cells. Thus there was evidence that there were activated T cells in the bloodstream of MS patients.

If VLA-4 was thought of as a "key," the next piece of the puzzle was to identify the "lock." In other words, what is the structure on the blood vessels to which VLA-4 fits that allows the T cell to bind to the blood vessel and enter the brain? This second piece of the puzzle was discovered by Roy Lobb and his team of scientists at Biogen in Cambridge, Massachusetts, who in 1989 and 1990 published two papers in the prestigious journal *Cell* in which they identified a molecule called VCAM-1 as the structure on blood vessels to which the VLA-4 bound. VCAM stands for "vascular cell adhesion molecule," which translates as "a structure on the blood vessel wall to which T cells adhere or bind." At the end of his article, Lobb wrote the following: "VCAM-1 may be central to recruitment of mononuclear cells (such as T cells) into inflammatory sites" in the body. This, of course, is one of the primary theories of MS, that T cells leave the bloodstream and cause inflammation in the brain and spinal cord.

Thus the basic science pieces were in place to perform an experiment in EAE, the animal model of MS. What has now become a classic EAE experiment was carried out by Ted Yednock who, at the time, was a scientist with a biotechnology firm called Athena Neurosciences in south San Francisco, California. Yednock decided to search for a monoclonal antibody that would block the VLA-4 structure on T cells and then test it in the EAE model. As we discussed in chapter 9, a Nobel Prize was awarded for the discovery of monoclonal antibodies, antibodies derived from a single clone of B cells, that all react with the same structure. If he found such a monoclonal antibody, it could then be adapted to be tested in MS patients. Yednock and his group tested the ability of a series of monoclonal antibodies to block the T cell from binding to blood vessels in the test tube. After lengthy testing, they found one that was very effective and thus had a monoclonal antibody in hand that they could test in the animal model of MS. It interacted with a structure called alpha-4-integrin, related to the VLA-4 structure that Hemler identified. The number of the antibody was HP2/1.

This is how they performed the experiment. First, they took T cells that were programmed to enter the brain of rats and cause disease and injected them into the bloodstream of the animals. They then injected the monoclonal antibody HP2/1 that attached to the structure on the surface of the T cell, and thus they blocked the ability of that T cell to bind to the blood vessel. As a control, they injected a different monoclonal antibody. What were the results? The animals that received the control antibody became paralyzed as the T cells freely entered the brain. The animals that received monoclonal antibody HP2/1 had no T cells in their brain and were not paralyzed. The experiment worked the first time. The results were published in the journal *Nature* in 1992, and the authors ended the article by saying that therapy directed at blocking the ability of T cells to enter the brain by using the monoclonal antibody could be effective in treating MS.

Now trying to block entry of T cells into the brain as a therapy for MS was a real possibility. Athena Neurosciences (which later became Elan Pharmaceuticals) found the results so compelling that they elected to proceed. However, the HP2/1 antibody was a mouse antibody. If one injects a mouse antibody into people, the people react against the mouse portion of the antibody; thus it could only be given to people a limited number of times. This is something we observed in 1986 when we injected a mouse monoclonal antibody into MS that reacted with T cells. However, in 1992 the technology existed to humanize a mouse monoclonal antibody so it could be easily given to people. Thus, in 1994 Elan humanized a monoclonal antibody from 10,000 that Yednock had screened (its number was AB4), and in 1995 Elan filed with the FDA to receive approval to test the antibody in MS. The antibody now was given a name; it was called *natalizumab.* All monoclonal antibodies have a name that ends in "mab," which stands for "monoclonal antibody," and the "u" prior to the "mab" means that it was a humanized form of the antibody.

The first study of natalizumab in MS was carried out in patients having an acute attack of MS. It was a placebo-controlled study. Patients in the midst of an MS attack were given an intravenous infusion of natalizumab or placebo. The goal of this study was to determine whether people that received the natalizumab recovered more quickly from their attacks. What were the results? Negative. In retrospect, based on the mechanism of action of natalizumab, treatment at the time of the attack would not have been expected to be efficacious. Why? Because when a patient had an attack, the T cells had already entered the brain. It was too late to block them. If attacks were caused by T cells entering the brain, it made more sense to treat prior to attacks and determine whether the attacks were lessened.

In order to ask this question in an expeditious fashion, scientists took advantage of MRI imaging. It is now well established that when T cells enter the brain and cause an attack in MS patients, they cause a breakdown of what is called "the blood brain barrier." In other words, there is free movement of cells and proteins into the brain that usually doesn't occur. This breakdown lasts for about four to six weeks after the T cells enter the brain and can be seen on MRI by injecting a substance called gadolinium into the vein and observing it leak into the brain by MRI. Thus a study was performed in which seventy-two patients with active relapsing-remitting MS and secondary progressive MS received two intravenous infusions of natalizumab or placebo given four weeks apart, after which the patients were then followed for twenty-four weeks both clinically and by MRI imaging. The results showed that the treated group exhibited fewer new active areas of inflammation on the MRI over the first twelve weeks. There was no difference in the second twelve weeks of the study. There was no change in the number of MS attacks with the treatment in the first twelve weeks; in fact, there were more attacks in the treated group in the second twelve weeks. The treatment, however, was well tolerated with no side effects, and the study was not designed to test relapses. The study was ultimately published in 1999 by

Tubridy, et al in the journal *Neurology.* The authors concluded that short-term monoclonal antibody therapy against VLA-4, now called alpha-4-beta-1-integrin, reduced new active areas of MS activity lesions on MRI, and that further studies were required to assess the clinical outcome.

Now that there was evidence that the treatment had an effect in MS as measured by MRI, Elan Pharmaceuticals was confronted with the decision of whether to proceed with a study in which treatment with the monoclonal antibody was given for a longer period of time. The answer was yes. A phase two clinical trial was launched that included 26 clinical centers in the United States, Canada, and the United Kingdom and enrolled 213 patients from September 1999 until May 2000. Patients received either 3 mg or 6 mg of natalizumab per kilogram body weight monthly for six months or a placebo. This was a major question in Ted Yednock's mind: Could natalizumab be given chronically and still maintain its effectiveness? The primary outcome measure was the number of gadolinium-enhancing lesions on MRI. If positive, a larger phase three trial would be undertaken that would have the potential to create a new drug for the treatment of MS.

The clinical development had proceeded far enough that it was now time to find a trade name for the new drug, which until now only had a number (AN100226). A contest was held at Elan, and Cathy Cannon, who worked in Ted Yednock's lab and was second author on the *Nature* paper that showed positive results in the EAE model, won $500 with the name *Antegren,* a name that described a drug that worked against an adhesion molecule.

The story of natalizumab, with its new trade name Antegren, then shifted to Biogen, the biotech company in Cambridge, Massachusetts, that manufactured and marketed Avonex, one of the interferons approved by the FDA for the treatment of relapsing-remitting MS. Although Biogen had major success with Avonex, they were looking to in-license another drug for the treatment of MS to broaden their MS drug portfolio. It was 2000, and the task fell to Nancy Simonian, who was senior director/vice president for clinical development. She was joined by Al Sandrock, a neurologist who came to Biogen in 1998 from the Massachusetts General Hospital, which was across the Charles River from Biogen.

As they looked at the different opportunities, the study of natalizumab by Tubridy caught their attention. It was far along in clinical development, having been tested in phase two studies, and some efficacy had already been demonstrated in patients. Furthermore, Biogen was very familiar with the scientific principles behind the drug, as they had their own program to develop both a monoclonal antibody and a small molecule against VLA-4. Roy Lobb at Biogen had made one of the key, basic observations of identifying VCAM as the structure to which VLA-4 bound. Simonian and Sandrock knew that a lot was riding on their recommendation, and they not only considered natalizumab, but other potential drug opportunities as well. Even after looking at other opportunities, natalizumab still seemed the most promising. But before making a final recommendation, Simonian requested the raw data from the Tubridy trial and put it in an Excel spreadsheet to examine it herself. She even blinded the data and sent it to Henry McFarland at the NIH for his opinion on whether there was a real biologic effect, and Rick Rudick, from the Cleveland Clinic, joined them in the strategy sessions. After all, Tubridy only gave two doses of the natalizumab and only observed a transient effect on MRI with no effect on relapses. They also reviewed data in which natalizumab was being tested in Crohn's disease, a disease in which there was inflammation in the bowel.

Sandrock felt that EAE was a good model for MS, especially as it related to how T cells migrated into the brain and caused damage, though he knew the risks of translating any treatment from animals into people. He was encouraged by the observation that people treated with natalizumab had an increase in the number of lymphocytes in the bloodstream. This demonstrated that natalizumab was having a clear biologic effect in people. Furthermore, Elan was interested in potentially partnering with Biogen, as Biogen had manufacturing capabilities.

Nonetheless, there were concerns about natalizumab. Was there a rebound when the drug was stopped? (There were increased attacks in the Tubridy trial in the second half of the study after the medication was stopped.) Would people make antibodies against the drug and become immune to it? Would the natalizumab cannibalize Avonex? Would it be efficacious and safe over the long term? What about the negative trial in acute MS? Simonian and Sandrock reasoned that if one day a drug cannibalized Avonex, it might as well be a Biogen product.

They were now under pressure to make a multimillion-dollar decision based on incomplete data. They knew that an Elan-sponsored, six-month double blind phase two trial of natalizumab, led by David Miller in London, was in progress. If they waited and the Miller trial turned out positive, the cost of acquiring the drug would skyrocket. If they invested in natalizumab and the Miller trial failed, they would have wasted the money. The fact that the acute study didn't work deterred others, but it didn't deter them. They believed natalizumab had a real biologic effect in MS. It was a drug with a new mode of action compared to current drugs on the market. In the end, they strongly recommended to the Board that Biogen enter into an agreement with Elan to bring natalizumab to the market. After deliberating, the Board at Biogen finally agreed, and a contract was signed with Elan to develop natalizumab for the treatment of MS. Now it came down to the simple matter of waiting for the results of the Miller trial.

Sandrock and Simonian were on the top floor of the Biogen building in Cambridge, Massachusetts, when the call came from Allison Hulme of Elan, who was in Texas where the Miller trial had just been unblinded. It was the holiday season, between Christmas and New Year's, and there were few people around at Biogen. With Sandrock writing, Alison began reading the number of enhancing areas in the brain in the treatment and control groups. There was a clear effect. Sandrock then asked for relapse data. "I was blown away," Sandrock said. "I was so excited. I didn't expect to see such a dramatic effect on relapse rate." For Sandrock and Simonian, it felt like a Christmas present and a clear vindication of their decision to partner with Elan in the development of natalizumab. For Sandrock, it was all that he had hoped for when he left academia, the ability to develop a drug that would help people with MS. After they presented the results to Burt Adelman, executive vice president of development, and to Jim Mullen, the CEO of Biogen, plans were immediately put into place to carry out a phase three trial and bring natalizumab (Antegren) to market.

The design of any phase three trial is crucial and involves strategic decisions that can make or break a drug. Furthermore, the landscape of MS therapeutics in 2001 was dramatically different than a decade earlier when there were no FDA-approved drugs for MS. When Simonian,

Sandrock, and Rudick sat down to strategize about a phase three trial, there were already five FDA-approved drugs on the market for MS, though the approved drugs were only partially effective. Thus, two trials were designed. The first repeated the standard MS trial design of drug versus placebo, whereas the second was an innovative trial design that tested the effect of natalizumab when added to patients on Avonex who had disease activity in the previous year, as measured by having an MS attack. Both trials were to run for two years, with endpoints consisting of MRI activity, MS relapses, and accumulation of disability. The FDA liked the combination therapy study more than the monotherapy study, as it was unique and provided an opportunity for a new drug that would help those not responding to conventional therapy with interferon. The concept of combination therapy had not yet been studied in detail in MS, though there were precedents in rheumatoid arthritis, another autoimmune disease. Most important, the FDA agreed that they would be willing to evaluate after patients had been treated for only one year and even consider potential approval of the drug based on one-year data.

Now that the phase three trials were designed, Sandrock needed someone to run the clinical trials. Careful administration of the trials was crucial if natalizumab was to be approved for the treatment of MS. Sandrock turned to Mike Panzara, a neurologist who was on staff at our Partners MS Center at the Brigham and Women's Hospital in Boston. Panzara had always wanted to pursue a career in MS, as someone very close to him suffered from the illness. He was actually working in Larry Steinman's lab when Ted Yednock's paper on the treatment of EAE was published in *Nature* and remembers the excitement at that time. Now he would have the opportunity to carry out the same experiment in people.

When Sandrock called, Panzara was in the process of setting up a major new initiative at our MS Center, in which we planned to track 1,000 MS patients over a ten-year period and follow them with blood testing, MRI imaging, and clinical exams. Although Panzara enjoyed his position at a leading academic center studying MS, the opportunity to head up a major phase three clinical trial that could bring a unique MS drug to market was compelling, and he was very impressed by the data Sandrock showed him. Thus, he elected to join Biogen in the summer of 2001 and run the phase three clinical trials for natalizumab.

Presentation of the Miller results and a meeting of investigators for the phase three trial were scheduled for an international MS meeting called ECTRIMS, held in Dublin, Ireland, in September 2001. As it turned out, the conference was held in the midst of the September 11 events in the United States; the impact of the Miller results was muted and planning for the phase three trials was delayed. (The Miller results were ultimately published in the *New England Journal of Medicine* in 2003.) An investigator's meeting was held a few months later in Orlando, Florida. Many people took the train or drove, as they were afraid to fly. A second meeting was held in Vienna. There were close to five hundred people at each of the two investigator's meetings, and Panzara remembers the excitement of the investigators as they considered the first new drug for MS in almost a decade.

The first big day for Panzara occurred in November 2001 when he received a call from a site in Berkeley, California, that the first patient was infused with natalizumab with no untoward side effects. The trial was underway. The trial in which patients received natalizumab or placebo was called the AFFIRM trial, and the one in which patients had natalizumab or placebo added to Avonex was called the SENTINEL trial. As is now the custom with all MS trials, a Data and Safety Monitoring Committee (DSMB) was established that included two neurologists (Fred Lublin and Jack Antel), a statistician, and a rheumatologist. Our center chose to participate in the SENTINEL trial, as we didn't feel we could treat our patients only with a placebo. Indeed, the majority of sites participating in the AFFIRM trial were overseas, a sign that it was becoming harder to recruit patients in the United States to a placebo arm in which patients received no treatment all.

The next challenge for every trial is to recruit the required number of study subjects in a timely fashion. The AFFIRM and SENTINEL trials were designed to enroll more than 1,000 patients each. Panzara thought that the AFFIRM trial would have more problems recruiting, as it had a no treatment arm, but it filled up more quickly than anyone expected, and in the end was actually overenrolled. At one point, 150 patients per week were enrolling. If anything, there appeared to be a lag in the enrollment for the SENTINEL trial. There were discussions of what to do, but with time enrollment picked up and soon all patients had been entered into the two trials.

Now all that could be done was to wait. For the next two years, Sandrock and Panzara and their partners at Elan watched as the two trials were underway. They could only hope that the monoclonal antibody they infused into the bloodstreams of the patients was binding to T cells, stopping the T cells from entering the brain, and in the end decreasing the number of new areas of disease activity on the MRI that would translate in a lower number of MS attacks that, in turn, would translate into less disability.

For Sandrock, the answer to the question came in late November 2003 when the trial results were unblinded and he got his first look at the data. He went with the statisticians to a tiny room at a hotel near Biogen, where he was shown the AFFIRM data, and then a week later to a different hotel to look at the SENTINEL data. "I couldn't believe my eyes," Sandrock told me. After one year of therapy, there was a dramatic decrease in both MRI activity and number of relapses. Even more remarkable, there was no appreciable toxicity. The results remained a closely guarded secret, and Sandrock did not want to unblind Panzara so Panzara could continue to run the trials unimpeded. Nonetheless, they immediately began discussions with Marc Walton at the FDA, and Sandrock began the arduous task of preparing a briefing package for the FDA. It was December. Both Sandrock and his boss, Burt Adelman, loved football and were New England Patriots fans. Sandrock had no time to watch any football games and had to fake football discussions with Adelman. If Adelman knew that Sandrock wasn't watching the games, he would have known something was up.

The Friday before Christmas, Panzara received a call from Sandrock, who told him that he was ready to share the unblinded results from the trial. "Mike," Sandrock said, "shut down your computer, take your bag, go to the Cambridge Marriott, and don't plan on going back to your office." When Mike entered a small conference room at the Marriott, he was given two folders with the code numbers for the AFFIRM and SENTINEL trials. In front of him was a flip chart, where Sandrock had written the number $p = 0.055$. As was the case for all drugs approved for MS, everything depends on the p value, the statistical level of significance of the results. If the results were not less than 0.05, they were not significant. Panzara was taken aback when he first saw the number 0.055, but quickly surmised it was probably a joke to make him nervous. Why else would he abruptly be called away from his office?

Panzara opened the folders to find that the p value for both trials was actually less than 0.001, with a 66 percent reduction in relapse rate in the AFFIRM trial and a 54 percent reduction in the SENTINEL trial. He immediately flipped to the safety data and found no significant toxicity or side effects. As they went through the data table by table, they now knew they had a chance for an early submission to the FDA and fast-track approval. He went home for the weekend and couldn't sleep. The only person he could talk to about the data was Al Sandrock.

January was spent preparing for an FDA prefiling meeting, and the CEOs of Biogen (now called Biogen Idec following a merger) and Elan were finally told of the results in early February, a week before the team headed to the FDA. The trip to the FDA was kept secret. When Sandrock was asked where he was going, he said he was going to Chicago for a meeting. The actual presentation to the FDA lasted but one hour, thirty minutes for data presentation and thirty minutes for discussion. They were given the go-ahead to file an application based on one-year data. The team was restrained during the presentation and ecstatic back at the hotel.

The presentation to the FDA was a material event for the companies and thus had to be made public. An announcement was made on February 18, 2004, that based on discussions with the FDA, Biogen Idec and Elan planned to submit natalizumab (Antegren) for FDA approval in mid 2004, based on one-year data. Panzara told me that the plan to protect the data was like something out of the CIA, and his greatest fear was that there would be leaks. There were no leaks, and he worked nonstop until the file was sent to the FDA in May 2004. The FDA then had six months to reply. The data contained tens of thousands of pieces of paper. In the past, trailer trucks were rented to deliver the boxes of paper to the FDA. Now it was submitted to the FDA on an electronic tape.

With the February announcement and the filing with the FDA in May, the MS community was abuzz. The availability of natalizumab to treat MS patients was now on a clearly defined timetable. The FDA, by law, had to give its verdict in November 2004, and the drug could become available for patients a few weeks thereafter. For Biogen Idec and Elan, this meant an enormous amount of work preparing the drug for manufacture, shipping, and marketing. The MS community was hoping to get its first look at the data October 9, 2004, in Vienna at the yearly European MS conference (ECTRIMS), but the FDA did not want the data shown yet. Thus, the potential excitement of seeing the data in Vienna dissipated. Nonetheless, other drug companies held meetings with MS experts to gauge the effect of natalizumab on the use of the interferons and glatiramer acetate in MS. I and many of my colleagues participated in these discussions, and it was a surreal experience. We were asked to comment on what the effect of natalizumab would be on prescribing other MS drugs without knowing the data. One unfortunate reality of drug development is that it is done by the private sector. The private sector is driven by investors, by Wall Street. Thus, MS specialists were bombarded by investors asking their opinion on what would happen after the approval and release of natalizumab.

The scientific community finally got to see the data on November 5 and November 20, 2004, when investigators who participated in the clinical trials of natalizumab gathered first in London, England, and then in Orlando, Florida, for the presentation of the one-year results of the data from both the AFFIRM and the SENTINEL trials. Susan Gauthier, from our Center, attended the meetings, as we had participated in the SENTINEL trial. Results were presented by Chris Polman and Rick Rudick, who were unblinded prior to the meetings. The results were greeted with great enthusiasm, and there was applause when the slides showing a 68 percent reduction in attacks were shown. The data was now there for everyone to see, and the frenzy of investors increased. Another issue that now confronted all doctors who treated MS patients was the need to give natalizumab by infusion. We had been treating MS patients with infusions of steroids and cyclophosphamide for many years, and thus we were well prepared at our MS Center, as we had an eight-bed infusion facility dedicated to MS patients.

At the meeting in Orlando where the results of the SENTINEL trial were presented, the last slide shown was the name under which natalizumab would be marketed. For years, the trade name of natalizumab was Antegren. At the last minute, the FDA asked that a different name be found, as it was felt that Antegren was too close to the name of a cardiac drug, and the FDA didn't want errors made at the pharmacy when natalizumab was prescribed. Thus Biogen Idec and Elan were told to come up with a new name, and the name *Tysabri* was unveiled. In the end, it didn't take long for doctors and patients to become familiar with the name Tysabri, though many that had worked for years on natalizumab still thought of it as Antegren.

Next came the big event, formal approval of natalizumab by the FDA, the ultimate culmination of the basic work by Lobb and of the experiments in the MS EAE model in rats by Yednock. The FDA was required to render a decision six months after the May 2004 filing, though they had the right to delay their decision by three months. Thus people at Biogen Idec knew the exact day approval was scheduled to come, and the marketing and shipping of natalizumab for patients was geared to this day. Two days before Thanksgiving, Al Sandrock found himself hovering around the fax machine waiting for the announcement from the FDA. Finally, at 5:30 p.m., the fax came. It was now official.

Natalizumab was the the sixth drug to receive approval by the FDA for the treatment of multiple sclerosis. What was unique about the approval of natalizumab compared to the other FDA-approved drugs was that it was approved based on one year's data. I spoke to Marc Walton about the approval of natalizumab at a meeting in Washington, D.C., about future clinical trials in MS. He told me it was the SENTINEL trial that led to one-year approval of natalizumab, as it offered to the patient another option should one of the interferons or glatiramer acetate not be efficacious.

The FDA approval of natalizumab was a major news event, and I found myself on television once again, this time standing in our infusion center. It was twenty-five years from the time I was on the national news when we reported positive effects of cyclophosphamide infusions for the treatment of rapidly worsening MS. There was a great deal of excitement for natalizumab, as it appeared to have stronger efficacy than the interferons or glatiramer acetate both on attack rates and shutting down inflammation on MRI as measured by gadolinum enhancement. Moreover, it did not have the toxicity associated with chemotherapy drugs. We were anxious to use the drug early in the disease to ask one of the crucial questions in MS: If one shut down inflammation in a complete fashion early in the disease, would it affect later degeneration?

Patients quickly took to the drug after its approval, and our first patient was treated in mid December 2004. By the end of January, we had treated more than a hundred patients, with many more scheduled. We began by treating those MS patients whose disease was not controlled on their current therapy or who were having difficulties with the injections. Some newly diagnosed patients that were just beginning treatment for their MS preferred a monthly infusion as opposed to injections, especially if it had the prospect of better efficacy. For those that were doing well on their MS drugs, we advised them to remain on their treatment.

I attended a national faculty meeting sponsored by Biogen Idec in Carlsbad, California, on February 15, 2005, where the current issues in MS were discussed. It was an exciting time. The meeting occurred over Super Bowl Sunday, and this time Al Sandrock was able to watch the New England Patriots win their third Super Bowl. Mike Panzara was ecstatic at the Carlsbad meeting. The reaction of the physicians told him that there was a large, unmet need for natalizumab.

With the launching of natalizumab, the treatment of MS was about to enter a new era. Natalizumab had a profound effect on inflammation in the brain. It worked by a different mechanism of action than the interferons and glatiramer acetate, even though the interferons were also thought to affect movement of T cells into the brain. It was given by monthly infusion as opposed to self-injection, and it didn't appear to have significant side effects. It was even more expensive than the other MS drugs ($20,000 per year), and were it to become widely used, it would become a multibillion-dollar drug that could significantly decrease the use of the interferons and glatiramer acetate. Furthermore, its use would affect the ability to test other drugs being developed for the treatment of MS. Questions remained, however. It was not clear for how long it could be given and whether side effects would develop over time. It had not been tested in progressive forms of MS. There was also the potential problem of patients making antibodies against the natalizumab, which would negate its efficacy, though this occurred in only 3 to 4 percent of patients taking the drug for two years. Of all the potential problems, however, what actually happened caught the medical world by surprise.

The week after returning from the national faculty meeting in Carlsbad, California, Mike Panzara was eating lunch when he received a message that there was a call from an investigator who had to speak to him urgently. Panzara took the call, only to learn that a patient that had been treated with natalizumab plus Avonex as part of the SENTINEL trial had died and was diagnosed with progressive multifocal leukoencephalopathy (PML), a rare neurologic disease seen most commonly in patients with a compromised immune system such as AIDS. Panzara immediately had the MRI and report faxed to him. As he was dealing with the news, hoping it was a false positive, a second case of PML in a patient treated in the SENTINEL trial was reported to him twenty-four hours later. He personally confirmed both of the cases. There was no doubt of the diagnosis. His only fear now was that there would be thirty more cases.

On Monday morning, February 28, 2005, almost a year from the time the code was first broken, and shortly after the Carlsbad meeting, Nancy Simonian was sitting in front of her computer at Millennium Pharmaceuticals talking to her nanny. An announcement flashed onto her screen that Biogen Idec stock had dropped precipitously, the company losing 43 percent of its market value and that Elan's market value had fallen 70 percent. She immediately opened the report to find that there had been a death associated with natalizumab, and it was voluntarily being taken off the market by Biogen Idec and Elan pending further investigation. Nancy was in utter disbelief, devastated. Even though she was now working with another company developing a different drug for MS, and Biogen Idec was a competitor, she felt as if she had lost a child. Everyone at Biogen Idec and Elan associated with the development of natalizumab had a similar reaction: confusion, denial, disbelief.

The suspension of the use of natalizumab pending a safety review set off another frenzy of calls from the news media, the investment community, and, of course, from MS patients. What did this mean for MS? Who else was at risk? What should patients do that had received treatment? Could it be prevented? When would the drug be reintroduced?

PML is a very rare disease caused by a virus called JC virus. Almost everyone is infected with the JC virus (80 percent of healthy adults), which lies dormant and rarely causes disease. PML most frequently occurs in people with a compromised immune system, particularly those with late-stage AIDS. It appears that by blocking T cells and their ability to get into the brain, natalizumab was blocking the surveillance function of T cells that kept the JC virus under control. It is also possible that it caused PML by somehow activating the virus in other parts of the body. The cases of PML caught everyone by surprise because there had been no increase in the incidence of infections in patients treated with natalizumab, including other types of viral infections. PML had never been reported before in patients with MS, and after review of the cases, the medical community concluded that PML was indeed related to treatment with natalizumab.

Questions were immediately raised as to whether the FDA had approved natalizumab too quickly. The two patients that came down with PML had been treated for thirty-eight and twenty-seven months with natalizumab plus Avonex. If the FDA had not approved natalizumab after one year and waited for two-year data, they would have known about the two cases, even though the PML occurred after the patients had already completed the two-year SENTINEL trial. In my discussions with Marc Walton, who headed the FDA committee that approved the drug, he emphasized how strong the data were and that few safety concerns had been identified, certainly not infection. Furthermore, in other autoimmune diseases such as rheumatoid arthritis, drugs were approved after six-month trials when it was clear that they affected inflammation in the joints. Walton also told me that approval of natalizumab was unique, as it was the first drug approved by the FDA that had been tested in combination with an already approved drug. As it turned out, it may have been that combination that led to the PML side effect of natalizumab.

On April 9, 2005, the American Academy of Neurology held its annual meeting in Miami, Florida. It is the largest meeting in the United States where new research on multiple sclerosis is presented. It was to be the first meeting in the era of natalizumab (Tysabri) approval, and there were many presentations on natalizumab that had been scheduled months earlier, prior to withdrawal of natalizumab from the market. Interest was so high that the hall where presentations on natalizumab were scheduled had to be changed to accommodate the extraordinarily large number of people. The final two-year results of the AFFIRM trial were presented, and not only showed a continued decrease in attack rate and MRI inflammation at two years, but a decrease in disability. Rudick presented the one-year results of the SENTINEL trial with two-year data scheduled for presentation in the fall European ECTRIMS MS meeting. There were additional presentations on MRI and immunologic effects of natalizumab, the use of natalizumab in combination with glatiramer acetate, tolerability, immunogenicity, and safety. The drug was extraordinarily safe apart from the cases of PML.

Russell Bart, an infectious disease specialist, participated in the session to answer questions about PML. As it turned out, there was now a third case of PML that occurred in a patient treated with natalizumab for the treatment of Crohn's disease, an inflammatory disease of the bowel, and the question of other cases in MS. On review of the cases, it became clear that in all instances, the patients had been on natalizumab plus another drug that affected the immune system, interferon in the case of the MS patients, and an oral immunosuppressant drug in the case of Crohn's disease. To date, there was no PML in patients that had only been treated with natalizumab.

The question now was whether natalizumab could be reintroduced for the treatment of MS patients. I spoke with Al Sandrock at the neurology meetings in Miami about this, and he told me that Biogen Idec and Elan were committed to doing everything possible to make natalizumab available to MS patients if after the full safety evaluation was complete, they believed that the benefits of natalizumab outweighed the risks. Furthermore, Biogen Idec is working on other drugs for the treatment of MS. In June 2005, the *New England Journal of Medicine* published the details of the three confirmed cases of PML, along with an editorial and an article about the current state of knowledge of PML. Sandrock, Panzara, and Adelman contributed a letter to the journal discussing the future use of natalizumab in MS, emphasizing that a better definition of the potential risks of its use will be key to understanding its place in therapeutics. In July 2005, I participated in an expert review panel in Boston hosted by Biogen Idec to review all the data that had accumulated and to assess the use of natalizumab to treat MS.

At this writing, it is not known what the future of natalizumab will be. It depends on whether other cases occur and how well patients can be monitored for toxicity. The MS community is hopeful that it can be reintroduced and used in a safe way. If no further cases are encountered, it could be reintroduced into the MS population, given as a single therapy, not in combination with interferons or other immunosuppressive drugs, with careful monitoring for toxicity.

Irrespective of how natalizumab is used in the future for the treatment of MS, what did we learn from the natalizumab experience? Most important, the underlying pathogenesis of MS and the ways in which it can be stopped were more firmly solidified. As Henry Claman said, "The more we understand about the disease and the more we can show that there are treatments that can affect the disease process in MS patients, the faster we will move toward ultimately curing the disease." The positive results with natalizumab thus moved us a big step farther toward ultimately curing MS. It clearly established the role of T cells and their migration to the brain as central to MS attacks, and that it was possible to shut down inflammation in a major way in the disease and that attacks were clearly linked to enhancing lesions on MRI. When I was asked to write an article in a series that dealt with controversies in MS that addressed the question as to whether MS was indeed a T cell autoimmune disease, the positive findings with natalizumab unequivocally established the role of T cells moving out of the bloodstream into the brain and causing damage.

Even though there are now six FDA-approved drugs for the treatment of MS, developing drugs for MS has always been a difficult process. When the two-month Tubridy trial of natalizumab was published in *Neurology* in August 1999, the same issue reported the failure of a phase two, placebo-controlled study of 168 patients treated with a drug that blocked the effects of a chemical called tumor necrosis factor or TNF. An editorial that accompanied the Tubridy study was entitled "Targeting Immunotherapy in Multiple Sclerosis: A Near Hit and a Clear Miss." The near hit referred to the natalizumab trial, and the clear miss referred to a trial of anti-TNF therapy that actually increased the number of MS attacks. TNF was felt to play an important role in the inflammation in MS, and drugs that blocked TNF (anti-TNF drugs) were successful in the animal model of MS, EAE. An interesting footnote to the TNF story is that anti-TNF therapy has proven effective in another autoimmune disease—rheumatoid arthritis—and has received FDA approval for the treatment of this disease. However, it has since been discovered that treatment of some rheumatoid arthritis patients with anti-TNF therapy unmasked latent multiple sclerosis in these patients. These findings underscore the complexity of the workings of the immune system in autoimmune diseases and that a careful balance must be maintained when therapy is given.

Natalizumab and anti-TNF illustrate two major hurdles that need to be overcome in the development of new drugs for MS. First, most drugs are first tested in animal models, and not all drugs that work in animals work in MS. Second, if a drug is found to work in MS, there may be toxicity or side effects that limit its use or makes it impossible to use in patients. There are other instances of drugs that were efficacious in clinical testing and whose development in MS was stopped because of serious side effects. One example is the drug linomide, which showed positive effects in MS but had unacceptable cardiac toxicity. In my discussions with Rick Rudick, we identified new questions for the MS community that have been raised by the natalizumab experience that are important for future MS therapeutics. For example, how much toxicity can the medical community accept with a highly effective drug, and who decides on its use? In this regard, combination therapy with more than one drug has been successful in the treatment of AIDS, and most experts feel combination therapy will be required for MS. Another question is how does the MS community move forward to test combination therapy in a way that will provide sufficient clinical information and that has an acceptable risk benefit?

Whatever its future, I view the story of natalizumab as a success story in our quest to find more effective treatments for MS and ultimately a cure. It introduced an innovative trial design that was the first large-scale trial of combination therapy in MS. Its mechanism of action on the immune system provides unequivocal support for pursuing the next generation of immune therapies. Marc Walton of the FDA told me that his tenure at the FDA has been an exciting one, in part because he has been there at a time when there has been a revolution in the development of FDA-approved drugs to treat MS. Even more exciting, there are scores of new drugs in the pipeline.

13

INFLAMMATION VERSUS DEGENERATION

WITH THE APPROVAL and widespread use of interferons and glatiramer acetate, a new era in the treatment of MS had begun. Now the neurologist had treatment options for MS patients and the approach was one of active intervention. MS had become a treatable disease. Nonetheless, as would have been predicted from the results of the large phase three trials, it soon became clear that the interferons and glatiramer acetate were only partially effective in MS. What did one do in patients who worsened despite therapy, patients who held out the hope that treatment would control their disease? One such person was Marilyn Bradshaw, a physician in training who was completing her internal medicine residency when I first began treating her for multiple sclerosis. Her husband was also a physician. She had been treated for four years with one of the injectable drugs by the time she finished her residency and left Boston for Chicago. I remember during her residency when I had her and her husband come to my house on an emergency basis because of new symptoms she was having. Being physicians, they were hypersensitive about her illness. Nonetheless, she was able to complete her residency and raise her two children. However, after taking a staff position in Chicago, her disease was not as controlled as it had been in Boston. She was having attacks that required treatment with steroids, and there were now emergency calls and even a visit to Boston because the attacks were interfering with her work as a junior faculty member and were beginning to leave some disability. After a long discussion with her and her husband and consultation with her doctor in Chicago, she was given a chemotherapy drug called mitoxantrone, which was approved in 2000, almost forty years from the time chemotherapy agents were first used in MS. Over the next two years her disease stabilized, and she was able to continue her medical work and raise her children.

The use of chemotherapy for the treatment of MS had haunted me since we published our results with cyclophosphamide in 1983. I knew the drug worked, as I had seen often dramatic effects with my own eyes in countless patients and we had positive results in our clinical trials. I also knew there were patients for whom it did not work. The controversy remained, as two controlled trials (especially one done in Canada) had not shown cyclophosphamide to have an effect. Over the years we struggled to understand the difference between our study and the Canadian study, hoping that it would lead to a better understanding of how to treat MS patients and, more important, a better understanding of the underlying mechanisms of the disease. The approval of mitoxantrone and clinical trials with the interferons in progressive MS led to an understanding of the differences.

The FDA approval of mitoxantrone was another major turning point in the therapy of MS, as it further validated the immune theory of MS and the approach of Gonsette and Hommes in the 1970s—the approach we took with cyclophosphamide in 1983. In addition to bringing the use of chemotherapy for the treatment of MS into the mainstream, it exposed what has now become the next major challenge in the understanding and therapy of MS—MS as an inflammatory disease versus MS as a degenerative disease.

For the FDA hearing on mitoxantrone on January 28, 2000, instead of sitting in the audience as I had during the FDA hearings for the interferons and glatiramer acetate, I now sat at the center table as a member of the advisory board. Dr. Sid Gilman, who had served as chair for the advisory panel on Copaxone, remained as chair, and Jerry Wolinsky joined me as an MS consultant to the panel. The hearing was also held in Gaithersburg, Maryland, and began with the usual introductions and conflict-of-interest statements. As he did for the Copaxone hearing, Dr. Russell Katz represented the FDA. As it turned out, it was the longest advisory committee meeting of all those held to approve an MS drug, most probably because of the complex nature of defining the clinical indications for the drug and the concern for its toxic side effects. Ironically, the data on the basis of which mitoxantrone was approved by the FDA for use in MS was not published in a peer-reviewed journal until December 2002, when the final report appeared in the journal *Lancet*. Thus it was almost three years from the time of the advisory board meeting before physicians could view all the data.

In the past, two well-controlled trials were required for a standard drug to be approved by the FDA. In November 1997 the law was amended so that only substantial evidence on a single controlled trial was required as long as there was confirmatory evidence. Russell Katz of the FDA began the proceedings by reviewing the two trials that were being presented to the FDA to support approval of the drug. The single controlled trial (study 1) on which the approval would primarily be based was a randomized, placebo-controlled trial of 194 patients in seventeen centers in four European countries that compared doses of 5 mg/m^2 and 12 mg/m^2 versus placebo given intravenously every three months for two years. The trial was led by Hans-Peter Hartung, a German neurologist. The primary outcome being measured was a complicated analysis that combined results of several disability scales. A typical patient was forty years old and had had MS for ten years. The second or confirmatory trial (study 2) was a five-center French trial led by neurologist Gilles Edan that had already been published in 1997; it involved forty-two patients, half of whom were treated with 20 mg of mitoxantrone plus intravenous steroids whereas the other half just received steroids. The primary outcome measure was based only on MRI results and consisted of the proportion of patients with no new gadolinium-enhancing areas on MRI. It was thus similar to our 1983 cyclophosphamide study in which we tested cyclophosphamide plus the steroid stimulating drug ACTH versus ACTH alone. For study 2, a different treatment regimen was used: monthly treatment for six months. Both Hartung and Edan were present and participated in the FDA hearing.

Study 1 enrolled either secondary progressive MS patients or relapsing progressive patients. Study 2 enrolled patients who had "severe" MS (a criterion Katz said was ill-defined) and "active" MS (which was defined as two relapses or an increase in disability in the year prior to enrollment). Katz stated that the proposed indication for mitoxantrone from the sponsor was "to slow progression of neurologic disability and reduce the relapse rate in patients with progressive MS." He noted that the two studies were performed outside the United States but that there was no requirement that trials had to be done in the United States for FDA approval.

A discussion then ensued regarding MRI, and Katz pointed out that MRI could not be the primary measure of effectiveness. Even though MRI effects had been crucial for the approval of Betaseron for MS, he did not know of an instance when CDER relied on a nonclinical measure to approve a drug. He did point out that cancer drugs were approved on the basis of tumor burden as measured by radiographic techniques, but warned that surrogate markers could be misleading. His reference to cancer was ironic as the advisory panel was considering a cancer drug, but for the treatment of MS. Although MS is not a cancer, it raised the conceptual point of considering the amount of damage to the brain as measured by MRI, analogous to the amount of disease a cancer patient had, and doing studies to reduce the damage. Furthermore, a common clinical outcome measure in cancer therapy is the five-year survival rate. One of the central clinical outcomes in MS that I felt was analogous to five-year survival rates in cancer was the ability to prevent patients from entering the progressive stage of the disease.

Katz ended his presentation by asking the panel to carefully consider the safety of the drug and, since it was a chemotherapy drug, whether its use should be restricted to oncologists or to neurologists in conjunction with oncologists.

It was time for the sponsor (Immunex Corporation of Seattle, Washington) to present its case, which was done by Dr. Richard Ghalie.

Mitoxantrone had been marketed under the trade name Novantrone in the United States and Canada since 1987 for a type of leukemia and prostate cancer. Over four hundred thousand people had been treated worldwide. It acted by inhibiting DNA synthesis and was a strong suppressor of the immune system. Like cyclophosphamide, it was very effective in the EAE animal model of MS. The major risk was that of heart disease. The sponsors had excluded treatment of patients with the primary progressive form of MS and those with benign MS.

Ghalie reported that all efforts were made to ensure that the trial was carried out in a double-blind fashion. Mitoxantrone is blue, and following treatment, the whites of people's eyes or their urine sometimes turned blue. To prevent this from being an indication of treatment, the control subjects (those receiving a placebo) were infused with a solution containing an inert dye called methylene blue that produced similar effects.

The results of both trials were reviewed by Ghalie, and both showed statistically significant effects on the outcome measures chosen, disability and MRI changes. Now it was time for questions by the advisory panel. One of the major questions raised was the exact type of MS for which the drug would be approved. If the drug was approved for patients whose MS was worsening, was it for patients who worsened because of more relapses or patients who worsened because of progressive disease? This question highlighted the gray areas that surrounded the clinical classification of MS. More important, it began to address MS as having both inflammatory and degenerative phases.

Thus one of the main questions discussed by the panel was the stage of MS in which the drug was actually tested and how clear the different categories of MS were. It was a thorny issue, not only for the panel but also for the MS community. Patients in the two trials of mitoxantrone being considered by the FDA included both relapsing-remitting and progressive patients. For entry, patients had to have had at least two relapses with accumulating disability or to be in the secondary progressive phase with rapid worsening on the disability scale. There was confusion about the different disease categories and how they blended into each other.

"Probably Dr. Lublin will help us with our confusion later," Jerry

Wolinsky said to the chairman, Sid Gilman, "but I think we are going to stay confused on this forever."

Fred Lublin, a neurologist who heads the MS center at Mt. Sinai Hospital in New York, then presented the results of a consensus survey on the clinical course of MS that had been published in 1996 with Stephen Reingold, research director of the National Multiple Sclerosis Society. An international survey of clinicians was carried out that led to a new set of standard definitions of the four most common clinical patterns of MS. The first pattern, relapsing-remitting, was the commonest form by which MS presented. The relapsing form was characterized by an acute flare-up of neurologic symptoms followed by recovery that was either complete or incomplete and thus could lead to accumulation of disability. Over time 50 percent of relapsing-remitting patients converted to the second clinical pattern, secondary progressive. In the secondary progressive form there was slow worsening without attacks, or if there were attacks, there was slow worsening between attacks. In the third form (the least common one), progressive relapsing, a patient began as progressive and then had superimposed relapses. The fourth clinical pattern was primary progressive, in which there was progression from the outset with no relapses.

Lublin said that apart from clinical patterns, there was no reliable way to distinguish the different forms, and aggressive therapy with a drug such as mitoxantrone would be indicated in relapsing-remitting patients who were undergoing stepwise deterioration, secondary progressive patients anywhere along the course, and those with the progressive relapsing form. There was no evidence that mitoxantrone helped in patients with the primary progressive form of disease.

I pointed out that there was a transition period for some patients when they went from the relapsing-remitting form into secondary progressive form, a stage often associated with inflammation in the brain as seen by gadolinium enhancement on MRI. Such patients were different from those who were slowly progressive, without enhancing areas seen on MRI, and could represent different categories related to the amount of inflammation in the brain. Clearly a major goal of treatment was to stop people from entering the progressive phase.

Jerry Wolinsky then asked whether we should treat a recently diagnosed person with mitoxantrone. This, in my mind, was a key question in the therapy of MS. If we really wanted to shut down the disease, shouldn't we give our strongest medication such as mitoxantrone or cyclophosphamide at the beginning, rather than wait for people to worsen? The answer: conceptually, yes; practically, no. For such an approach we needed to know which patients were at greatest risk for worsening and we needed drugs as strong as mitoxantrone and cyclophosphamide without the side effects.

I followed up by asking whether, based on our experience with cyclophosphamide, there was less of a response to mitoxantrone in patients who had been in the progressive phase for over two years. We were finally beginning to understand which patients responded to cyclophosphamide, and one would assume that the same would be true for mitoxantrone. Those who responded to cyclophosphamide were those in the inflammatory stage as opposed to the degenerative stage of the illness—those with attacks and inflammation on the MRI as measured by gadolinium enhancement. Furthermore, those who only recently began the progressive phase responded better and patients with primary progressive MS did not respond well. Dr. Ghalie answered that the data needed to answer this question were not available, but he told me he had expected me to ask the question.

Nonetheless, there were data from the study of interferons in progressive MS that supported the hypothesis that treatment of later stages of progressive MS with anti-inflammatory drugs such as interferons or cyclophosphamide was not as effective in slowing the accumulation of disability. A European trial of interferon beta-1b in secondary progressive MS was effective in slowing the accumulation of disability, whereas a North American trial using the same drug was not, even though both trials showed an effect on the MRI scan. What was the difference between the two trials? As it turned out, patients in the European trial were younger and less likely to be free of exacerbations in the two years before the study. In other words, it appeared that the reason for the difference may have been that there was a stronger inflammatory component to the illness of those in the European trial. This had important implications for the use of a drug such as mitoxantrone or cyclophosphamide, namely, that the later one was treated in the course of progressive MS, the less likely it would be effective.

The discussion then led to MRI imaging and how it related to the clinical course of MS, since in one of the trials being considered, the number of enhancing areas on MRI imaging was the outcome being tested. I reported on an MRI study we had conducted with Charles Guttmann in which we performed twenty-two MRIs on a single patient in one year, including a period of weekly MRIs. That study yielded a dynamic view of the brain, and we had a time-lapse video made to demonstrate that MS could be very active as viewed by MRI even if the patient was stable clinically. It also showed a relationship between gadolinium-enhancing areas and numbers of attacks a person had, and a relationship between the accumulation of MS plaque in the brain and disability. Nonetheless, our studies and others that investigated the relationship between MRI and disability in MS found that MRI did not correlate strongly with disability. Although this was a major conundrum in the use of MRI to study MS, someone on the panel said, "Growing white spots on your brain is inherently a bad thing," irrespective of its exact correlation with disability in short-term studies.

We do know some of the reasons that the correlation between MRI and disability is not perfect. First, there are "silent" areas in the brain and if a patient had an MS plaque in a silent area, there probably wouldn't be any symptoms. Second, the brain can compensate for MS plaques by using other areas to carry out brain function. Third, our main measure of disability was motor function, and Lublin pointed out that some of the findings on MRI may correlate better with cognitive function, as measured by neuropsychological testing, rather than walking. Nonetheless, since MS is a chronic disease, it was discovered that when patients were followed over a longer period of time there was a better correlation between clinical course and MRI. As discussed previously, this was best demonstrated by a fourteen-year study from David Miller's MRI unit at Queen Square in London, which showed that patients with a larger burden of MS on MRI when first studied did worse when they were evaluated fourteen years later.

Furthermore, it was becoming clear as more was learned about MRI and the MS process that damage to axons (nerve fibers) was more important in causing long-term disability than the inflammation measured by gadolinium enhancement, though the first was involved in triggering the second. This difference highlighted the difference between the inflammatory and degenerative aspects of the disease. One of the reasons conventional MRI imaging does not correlate that strongly with disability is that it is not a sensitive measure of degeneration. Special techniques are required to measure the nerve damage caused by degeneration and correlate better with disability. One such technique is MRI spectroscopy, which measures chemicals in the brain, including NAA (n-acetylaspartate), which is found in axons. Doug Arnold, a neurologist at McGill University, has shown that NAA levels are decreased in MS and are an important measure of nervous system damage. Another technique, MTR (magnetization transfer ratio) NAA, has been shown by Massimo Filippi and colleagues in Milan to demonstrate tissue damage in normal-appearing white matter that cannot be seen on conventional imaging, telling us that MS is more widespread in the brain than we thought. Of note, however, is that although there were instances in clinical trials of positive effects on MRI with no clinical effects on MS, there were no trials with positive clinical effects in the absence of positive effects on MRI.

Everyone agreed that what was needed in MS was a surrogate marker of the disease process—a marker that predicted the clinical outcome. An example is the use of blood pressure as a surrogate for heart disease. Could MRI be used as a surrogate marker in MS? Experts in the field addressed this issue at two international meetings, one in Oxford, England, in 1997 and one in Washington, D.C., in 1999. It was agreed that MRI could be used as a surrogate for pilot trials of new drugs and for phase two trials to check if a drug was worth studying further, but it was not ready yet for phase three trials that would lead to approval of a drug. One day, it was hoped, a surrogate measure such as MRI could be used to approve a new drug for MS in a phase three trial.

It was time for a lunch break, and Gilman advised the panel not to discuss the drug under consideration at lunch. All discussion should be in public.

After lunch, the hearings resumed with patient presentations. A man with secondary progressive MS spoke and told the panel that taking interferon or glatiramer acetate didn't help him and he had heard that the mitoxantrone was specifically targeted for secondary progressive MS. A woman with five-month-old twins also told of the need for additional therapy when the injectable treatments were not effective. She said she wanted something for her MS if it changed to the secondary progressive form. Another person who had accumulated significant disability over twenty-five years told of her battle with MS and not wanting others to suffer like she did.

After hearing the patient presentations, I cautioned that we must be careful about creating false hopes among the severely disabled or those with only slow progression, as mitoxantrone was not likely to help much in those with later stages of progressive MS.

Safety data was then reviewed: the potential for cardiac problems, the effect on white blood counts, the effect on pregnancy. The committee wanted an absolute cap at a dose of 140 mg/m^2 to prevent cardiac side effects. Other side effects included loss of menstrual periods in older patients, nausea or vomiting, and hair thinning.

After a 4 o'clock break, there were more questions. Because the trials included relapsing-remitting as well as secondary progressive patients, Katz again brought up the issue of whether relapsing-remitting and chronic progressive MS constituted a continuum of disease or were distinct disease entities, both in the underlying mechanisms of disease and in the response to treatment. I pointed out that the immune response was different in the two forms and the MRI patterns were also different. I felt that while the disease was transitioning from relapsing-remitting to progressive, it was part of a continuum, but after it had been progressive for a period of time, there were enough differences for it to be considered distinct, a line was crossed, and as in the funeral joke, it "became funny" as quantitative changes became qualitative.

After further discussion on the role of MRI, surrogate markers, how well the trials were blinded, and dosing, it was time to vote. Because of the confusion over the relationship between relapsing-remitting and progressive forms of the disease, there was even more discussion on the wording of the indications for the drug. It was decided to vote on mitoxantrone for effectiveness in "clinically worsening, secondary progressive or relapsing remitting multiple sclerosis," with final details of the wording to be worked out with the sponsor.

Finally, at 5:27 p.m., mitoxantrone was approved unanimously by the advisory panel.

The approval of mitoxantrone closed a circle that was needed for the treatment of MS even if the drug was imperfect because of its potential toxic effects. It legitimized the use of strong suppression of the immune system for the treatment of MS and provided something for patients not helped by the injectable therapies. Doctors now had drugs to give at the beginning of the disease and treatment for when the disease was worsening. Furthermore, its approval heralded the beginning of a new phase of MS treatment and clinical trials.

This new phase was brought into focus at a conference in Seville dedicated to the use of immunosuppressant drugs for the treatment of MS. Gonsette and Hommes, who had pioneered the use of cyclophosphamide, were present, and Gonsette introduced mitoxantrone. I felt we were finally making progress in our understanding of how these drugs worked and when in the disease they were the most effective. In addition, a new category of clinical trials was being carried out in patients who continued to have attacks and worsen despite injectable therapy, and the concept of "rescue therapy" was born. We carried out a rescue therapy–type trial using cyclophosphamide in patients classified as nonresponder patients, people that continued to have attacks or worsen while on injectable therapy. The trial design was similar to the one we used when we first published our results on cyclophosphamide in 1983 and to that of the MRI-based mitoxantrone trial submitted to the FDA. Patients who were not responding to injectable therapy were randomized to receive six months of treatment with outpatient infusions of the steroid methylprednisolone or with the steroid plus cyclophosphamide. We found that those that received cyclophosphamide had improvement on the MRI and reduction in the number of attacks compared to those that only received the steroid. After they had stabilized on the cyclophosphamide, we put them back on injectable therapy, which, unlike cyclophosphamide, could be given over longer periods of time. The question was how they would do over time and whether six months of treatment with cyclophosphamide was enough. Like mitoxantrone, cyclophosphamide is used as rescue therapy in worsening MS, though no drug company has sought formal FDA approval for its use in MS since it is no longer on patent.

After the approval of mitoxantrone, some proposed even stronger immune suppression using bone marrow transplantation or very high doses of cyclophosphamide. Nonetheless, the future of strong suppression of the immune system in multiple sclerosis lay not in drugs such as mitoxantrone and cyclophosphamide but in the development of a new generation of drugs that modulated the immune system in a specific fashion. Robert Schwartz, an expert in immunology, wrote about this class of drugs in an editorial entitled, "The New Immunology—the End of Immunosuppressive Drug Therapy?" which accompanied the report of the successful use of one of these drugs in transplantation. Such drugs, many of them monoclonal antibodies, already have been shown to be effective in the EAE animal model of MS and are being tested in initial trials in MS.

At the FDA hearing for mitoxantrone, Jerry Wolinsky had raised the question of whether mitoxantrone should be given to everyone when their MS was diagnosed. It was a theoretical question, as mitoxantrone was too toxic for such use, but this was not the case for the new immune modulators, and their use at the onset of MS might one day have an enormous impact on the disease.

ALTHOUGH THE approval of mitoxantrone for worsening MS was an important milestone in MS therapy, it also served to highlight the next major challenge in our understanding and treatment of MS. It was becoming increasingly clear that in addition to the inflammation that occurred in MS, there were other processes associated with worsening in MS and thus new treatment approaches that were different in their mechanism of action from mitoxantrone and cyclophosphamide were needed. The scientific paper that brought this into focus was published by Bruce Trapp in 1998. Trapp, a scientist at the Cleveland Clinic, reported that not only was there loss of myelin in multiple sclerosis, but that the nerve fibers or the axons were also affected. In fact, they were transected or cut by the inflammation. He published his paper in the *New England Journal of Medicine* along with beautiful color pictures of the nerve fibers that had contracted into green balls after they were transected. His paper highlighted a pathological finding that had been described more than a century earlier by the French neurologist Charcot. Trapp's report rekindled a great deal of interest and forced the reevaluation of the processes at work in the brains of MS patients that led to disability.

On evaluating his findings, some pathologists said that we already knew that nerve damage occurs in MS and that Trapp hadn't discovered something new. Nonetheless, all scientists build on what other scientists have discovered previously, and Trapp's report led to a new focus on why patients with MS worsened over time. Those suffering from MS do not really care who was the first to make a discovery. They care only that scientists are focused on the right question, the answers to which will stop the worsening of the disease and in the end lead to a cure. This becomes most obvious to scientists or clinical researchers when someone in their own family contracts MS.

The findings of Trapp highlighted what has now become a major research and clinical theme in MS: "inflammation vs. degeneration," and related to the use of drugs such as mitoxantrone in progressive MS. Although we believe that inflammation initiated by T cells targeted to myelin proteins in the brain cause MS, it is now clear that the inflammation may then trigger degenerative changes that are independent of the inflammation. In other words, therapy such as mitoxantrone and cyclophosphamide designed to stop T cells from damaging the nervous system may not be as effective once the damage is done. Furthermore, the damage to the nerve fibers or axons begins early in the disease process, and there is a relationship between axon damage and the degree of inflammation in active MS areas of the brain.

The components of inflammation versus degeneration in MS are highlighted by clinical studies carried out by Alastair Compston, an English neurologist and MS expert, using a monoclonal antibody called CamPath. CamPath, named for its development in the pathology department at the University of Cambridge, England, is a monoclonal antibody that acts as a general immunosuppressant that is targeted to a molecule on all lymphocytes and is very effective in shutting down inflammation in MS as measured by clinical attacks and gadolinium enhancement on MRI imaging. What Compston and his group reported is that even though CamPath helped the MS process by shutting down inflammation, it did not affect the underlying degeneration triggered by the inflammation. Thus, the improvement patients experienced following CamPath as measured by decreased number of relapses was accompanied in some patients by a slow worsening, presumed secondary to the degenerative process.

The issue of degeneration triggered by inflammation raises important research questions. How do we study the degenerative processes in MS in animals? How can we test new therapies? In my discussions with one of the MS fellows at our center, this exact question came up: "If MS has a major degenerative component, especially in the later stages of the disease, does that mean EAE, which is an inflammatory disease triggered by T cells, is not a good animal model for the degenerative stages of the disease?" The answer is no. As scientists have continued to study the EAE model, it appears to mimic MS not only in its inflammatory components but also in its degenerative components. In EAE, degeneration follows inflammation just as it does in MS patients. Thus there are EAE models in which animals enter a chronic progressive phase of an illness just like the MS patient. Furthermore, Cedric Raine, an expert MS neuropathologist, has shown that drugs that do not affect the immune system but which block toxic substances, such as glutamate, that are associated with degeneration can lessen nerve damage and the degree of paralysis in the EAE model. A new class of drugs that are designed to block the degenerative process are now being developed, and testing has already begun in some MS patients.

For the individual MS patient, it appears that a line is crossed when the patient changes from a relapsing-remitting pattern to a progressive one. Studies using a large European database headed by Christian Confavreux have shown that patients with the relapsing-remitting pattern of MS can take from one to thirty-three years to reach a disability level of 4, but once they reach that disability level, which is associated with entering the progressive phase, it takes between four and seven years to reach a disability level of 7 (wheelchair-bound). These findings suggest that it may be damage to the axon that is the main cause of irreversible damage during the progressive stage of MS. Thus it appears that there can be silent damage occurring that the patient does not feel and when the threshold is crossed, it may be too late for treatment with currently available drugs. This argues for early therapy of MS patients (point eighteen of the twenty-one points).

Why, then, do some patients do well for many years even though they have attacks? One explanation is that the brain is able to compensate for the damage caused by attacks, but when a certain threshold is reached, the brain can no longer compensate. In support of this is a study in which functional imagining of the brain was carried out in patients with relapsing-remitting MS who had minimal disability. In functional imaging, different parts of the brain light up when a person performs a task. When patients with MS were asked to move their hands, there was a fivefold increase in brain activity compared to individuals without MS. Thus in MS patients the brain is working overtime to keep things functioning normally. As more damage is done, the brain is no longer able to keep up, and the hidden damage becomes clinically apparent and results in disability. Nonetheless, there are some patients with relatively benign disease who don't accumulate disability over time. It is possible that such patients have factors that help protect their nervous system from the toxic substances or the inflammation does not induce a high level of toxic substances in their brains.

Although the appreciation of degenerative changes that occur in MS means that we lack treatments for certain stages or types of MS, the knowledge is encouraging. A better understanding of the degenerative changes will lead to development of drugs to prevent axon damage and will ultimately impact the disability from the progressive forms of the disease. In the future we may be treating MS patients with drugs that control degeneration in addition to those that target inflammation.

In my discussions with Alastair Compston and Bruce Trapp, we have explored the interrelationship between inflammation and degeneration. Some even argue that MS is a degenerative disease and that all inflammation is secondary. I pointed out that in progressive forms of the disease, not only does the nervous system change, the immune system may also become chronically active as well. "What is the big experiment to do?" I asked Alastair.

"I think it is pretty clear," he said. "Treat aggressively in early stages of the disease to shut down inflammation, and then we will see if the degeneration doesn't occur."

14

TWENTY-ONE POINTS

For people of my generation, everybody remembers where they were when they first heard that John F. Kennedy was shot. I was a student at Dartmouth College walking up the second flight of stairs to attend a history lecture on the third floor. I think of that moment from time to time when I walk in my hometown of Brookline, Massachusetts, and pass the street where Kennedy was born. Now there is a second date, September 11, 2001. On that day I was in Belfast, Ireland, checking into the Europa Hotel, an hour and a half before I was scheduled to give a lecture to the MS Society of Great Britain and Northern Ireland. Someone later told me that the Europa Hotel was one of the most bombed hotels in Europe. I had just been driven in a tiny car from an MS meeting I was attending in Kildare, Ireland. As I went up in the elevator, the porter said to me, "Dr. Weiner, have you heard what happened in the United States? You better turn on the television." I did, and I couldn't believe my eyes. *How am I going to give my talk?* I thought.

We drove to the lecture hall, which happened to be adjacent to a golf course. I briefly looked out over one of the fairways, the green pastoral scene offering a moment of solace. I later read an article by Thomas Friedman of the *New York Times* about how in Beirut people had played golf, even in the midst of war, defying the interruption of normal life.

I was invited to give the lecture by Dr. Lorna Layward, who was the director of research for the MS Society of Great Britain and Northern Ireland. We had discussed what would be the best topic for my lecture. I would be speaking to a predominantly lay audience of MS patients and their families, with a sprinkling of doctors and one or two MS specialists. It was a named lecture called the "Jack Pritchard" lectureship in honor of one of the prominent neurologists in Northern Ireland. After the lecture, I was given a large letter opener as a memento of my talk with my signature and the date September 11, 2001, carved into it, an indelible reminder of the day and of my talk.

"What do people want to hear?" I asked Dr. Layward after I accepted the invitation.

"The usual," she said, "treatments, research, what's new in MS. You should know that people are very excited about your coming. We are expecting almost three hundred people. Try to do something special."

I thought for a second. What could I do that was special? If they *really* wanted to hear about MS, why not give them the twenty-one-point talk? It was a talk that I had given countless times, but always to scientists and in scientific terms. Why not give it to a lay audience and try to translate it into lay terms?

"Can you do it?" Dr. Layward asked when I told her of my idea.

"I'll try," I said, and was actually looking forward to the challenge. I decided I would show the exact slides I showed in scientific lectures but would translate and try to involve the audience. The biology of a disease was not physics or mathematics, one had to be able to explain it to nonscientists.

I walked up a steep flight of stairs from a small room in the basement where we were watching the September 11 events on television, and entered the lecture hall. The lecture room was packed. I discovered they were unable to project my slides onto the screen. Fortunately, at the last minute, the problem of my PowerPoint presentation was solved. After words of condolences from those that introduced me about what had happened in the United States and my acknowledgment of the September 11 events, I began.

I put up the first slide. It was the same slide I used for my scientific talk and it read as follows:

> *1. MS is a cell-mediated autoimmune disease directed against myelin antigens such as MBP, PLP, and MOG. Autoantibodies may play a secondary or enhancing role.*

I read the slide for the audience and then said, "Don't be intimidated by the slide; I will translate. The first point forms the basis of what we believe causes MS," I said. "What causes MS? Is it a toxic substance in the body? Is it a virus? No. Point one says that the factor that causes MS is a white blood cell called a T cell. It is a T cell that moves out of the bloodstream, goes to the brain, and damages the myelin sheath, the covering of the axon, or nerve fiber. Because it comes from the thymus, which is a gland in the neck, it is called a T cell. People from New York are called New Yorkers, people from Dublin are called Dubliners. White blood cells or lymphocytes that come from the thymus are called T cells.

"The T cell is the primary cell that orchestrates our immune system. A T cell helps fight off infection, and it upregulates and downregulates the immune system. An easy way to think about how a T causes damage in MS is to think of transplantation." I turned and pointed to Professor Ingrid Allen, who had introduced me. "If I were to transplant a kidney from Professor Ingrid Allen into myself," I said, "my body would reject her kidney, even though I am fond of her." People laughed. "How does my body reject a kidney transplant if it isn't from an identical twin? It is done by T cells. T cells recognize that the kidney isn't from my body and they attack it. This is what we believe happens in multiple sclerosis, only in the case of MS the T cell attacks one's own tissue. This is why MS is called an autoimmune disease. The immune system attacks one's own body. In MS, for whatever reason, the T cell thinks that the myelin in the brain and spinal cord are foreign, like a foreign kidney transplant, and they attack it. MS isn't the only autoimmune disease—juvenile diabetes, rheumatoid arthritis, and certain types of thyroid disease all involve a T cell gone astray, a T cell attacking one's own body. Under the microscope, we can see T cells and other white blood cells in the brains of MS patients doing their damage.

"Now, what do MBP, PLP, and MOG stand for?" I asked, pointing to the slide. "These are abbreviations for the structures or proteins in the brain that the T cells attack. Remember, the T cells only attack the brain and spinal cord; they don't attack any other part of the body, not the liver, not the kidney, not the joints. T cells are specific. They are programmed for only one target and can go nowhere else. Our belief that T cells are the primary cause of MS was established by finding such myelin-reactive T cells in MS patients and by showing in animals that if you create these T cells they can cause a disease similar to MS."

I then pointed to the second sentence: *Autoantibodies may play a secondary or enhancing role.* "There are two basic parts of the immune system: cells and antibodies. Although we believe that antibodies as well as T cells may play a role in MS, the primary culprit is the T cell. Without abnormal T cells there would be no MS.

"So we have point number one," I said. "MS is caused by a T cell gone astray."

I then put up the next slide.

> *2. There is no single autoantigen in MS because after an immune attack on one myelin antigen, there is spreading of reactivity to other myelin antigens.*

"The second point is simple," I said. "The T cell can attack more than one target in the brain. In fact, it may begin by attacking one target in the brain and then different T cells attack another target in the brain. This point has major implications for trying to treat MS by eliminating specific T cells. When I first began studying MS in 1971 one of the central unanswered questions involved the specific target or structure in the brain that was being attacked by T cells. For almost twenty years the holy grail in MS was trying to identify the unique structure in the brain that was being attacked. In fact, Jonas Salk tried to cure MS by desensitizing or shutting down T cells to MBP, one of the brain proteins. However, we now know that there is more than one target that T cells attack in the brain. There may be as many as ten different targets that the T cell attacks. This raises the question, which we will discuss later, of whether there are different subtypes of MS, depending on which brain target is attacked. Furthermore, if a T cell attacks one brain protein, other structures are damaged, and they can sensitize additional T cells to attack other targets. The scientific term for this phenomenon is *epitope spreading.* Thus trying to find the single target being attacked by T cells is no longer relevant. Unfortunately, this makes design of specific therapy complex.

"To review," I said, "point one says that T cells targeted to brain structures cause MS. Point two says they may be directed against more than one target in the brain, and after they attack one target, they may become triggered to attack other targets.

I stopped and took a drink of water. "The next question is obvious. If MS is caused by a T cell that attacks the brain but not everyone gets MS, what triggers the T cells to go into the brain?" I put up the third slide.

> 3. *Initial sensitization is secondary to cross-reactivity between infectious agents and CNS myelin or self-limited infection of the brain that releases myelin antigens.*

I looked out at the audience and asked a question. "Why do we have an immune system?" No one answered. People thought it was a rhetorical question, but I decided it was time for me to involve the audience. I asked the question again, and this time pointed to a man sitting in the first row.

"To fight off infection," the man said finally when I kept pointing at him.

"Exactly," I answered. "And is there ever a time when the immune system fights off an infection and causes damage in the process?" No one answered the question and I didn't expect an answer. "What about getting a strep throat?" I asked. "Is it good or bad to get a strep throat?"

"Bad," someone in the audience finally called out.

"If you get strep, what do you get treated with?" I asked.

"Penicillin," someone else answered.

"Exactly," I said. "Strep throat is caused by an infection—by a bacterium called streptococcus. That is why it is called strep throat. If one gets a strep infection, the body's immune system attacks the bacteria and knocks it out. There is only one problem. There are structures on the surface of the bacteria that are also found on the heart, and if someone has a strep throat and the immune system is given a chance to attack the bacteria, it can also attack the heart. This is what causes rheumatic fever, which damages the mitral valve of the heart. By treating immediately with penicillin, the bacteria are killed and the immune system doesn't have time to get activated.

"We now know that there are many viruses and bacteria that have structures similar to myelin proteins in the brain. When one has an infection, the immune system attacks the virus, but because there also is a structure similar to the virus in the brain, the immune system then attacks the brain thinking that it is the virus. In scientific terms this is called *molecular mimicry,* in common terms, 'a case of shared identity.' Thus a major theory to explain how T cells which attack the brain get triggered is that during the course of routine viral infections T cells are accidentally triggered to attack the brain because there are structures on viruses similar to those in the brain. Another possibility is that in some instances there may be a self-limited infection in the brain, just like one gets a respiratory or intestinal infection, and when that happens the infection in the brain releases brain proteins that then trigger the T cells. In either case, the T cells are triggered because of infections."

The fourth point led to one of the favorite questions I liked to ask, whether I lectured to scientists or to lay audiences, and has become central to our understanding of T cells and MS. Thus, before putting up the fourth slide, I asked, "I have a question that can be answered yes or no," I said, "and I want everybody to answer by raising their hand. The question is simple. If we believe that MS is caused by T cells that attack myelin structures in the brain, there are two possibilities: either only people who have MS have T cells that can cause the disease, or everyone has T cells that can cause MS but the cells are only activated in MS patients. Remember, a major question is, Why doesn't everybody get MS?"

The question caused a murmur in the audience and I glanced at one of the scientists who was smiling because he obviously knew the answer.

"How many believe only MS patients have T cells in their bloodstream capable of causing MS?"

People looked at each other, murmured, and began to raise their hands. Approximately a third raised their hands.

"How many think that we all have T cells in our blood that can cause MS?" I asked.

Another third of the audience raised their hands.

"It is clear that some people haven't voted," I said with a laugh.

"The answer is that *everyone* has T cells in their blood capable of causing MS," I said. "Now, the big question is, why doesn't everyone get MS? Why do these T cells cause MS in some patients and not in others?"

Now I was ready to put up the slide that had the fourth point.

> *4. Autoreactive T cells for myelin components exist in normal individuals. The major determinant of disease induction is the class of immune response that occurs when these autoreactive T cells are triggered in MS patients.*

"The fourth point, although oversimplified, is as easy as one-two-three," I said. "We believe there are different classes of T cells that recognize structures in the brain. One class is called Th1 cells. When triggered, the Th1 cells cause damage and lead to MS. The other classes are called Th2 and Th3 cells, and if they are triggered they are harmless. More than that, Th2 and Th3 cells are regulatory cells that can keep Th1 cells under control and there are other types of regulatory cells as well. In animals one can cause an MS-like disease by generating Th1 cells that are targeted to brain proteins, and one can suppress disease by generating Th2 and Th3 cells directed against the identical brain proteins. Factors that favor the development of Th1 cells are elevated in MS patients and are also triggered by viral infections. One of them is called gamma interferon; another is called IL-12. Later when we discuss treatments that affect the immune system and help MS, we will find that almost all of them decrease Th1 responses and increase Th2 and Th3 responses. The concept of Th1, Th2, and Th3 cells has only been developed in the last ten years. When I first began studying MS, many researchers did not believe that normal people had cells that could attack the brain. One of the first demonstrations came from the laboratory of Hartmut Werkele, who found brain-reactive cells in normal rats that could induce MS disease in animals. Th3 cells were actually discovered in our own laboratory."

I then showed an animal experiment. Two different inbred strains of animals were injected with the identical myelin protein. One group came down with an MS-like disease but the other group did not. "What is the difference between these two strains of animals?" I asked. "Why did one group of animals get MS and the other did not?"

A number of people were ready with the answer. A woman in the back row called out, "The animals that got sick had a Th1 response and the animals that didn't had a Th2 or Th3 response."

"Go to the head of the class," I said. "Remember, there are always exceptions and biology is more complex than one-two-three, but the Th1-Th2-Th3 paradigm appears to be basically right. It also provides us an avenue by which we can understand why some people may get MS and others don't, and a framework for therapy: we want to decrease Th1 responses and increase Th2 and Th3 responses. Furthermore, the study of MS involves understanding the factors that can push immune responses toward Th1 versus Th2 and Th3. As I mentioned, there are also other types of regulatory T cells that the immune system can use to shut down Th1 cells (CD25 + T cells, CD8 T cells, NK T cells). Unfortunately, there are many factors that can influence the balance, and this is what makes understanding MS so complex.

"Are you ready for the fifth point?" I asked. After a few scattered yes's, I put up the next slide. I thought back to when I gave the twenty-one-point talk to a group of neurologists in Mumbai, India. One of them said to me afterward, "Dr. Weiner, when you hit twenty-one I was sad because I wanted more points."

The fifth point stated that genetic influences determined which brain proteins an MS patient's T cells reacted against, and they also determined the class of immune response that occurred.

> 5. *Generation of pathogenic autoreactive T cells is favored both by major histocompatability complex (MHC) and non-MHC genes, which determine which protein sequences an individual reacts against and the class of the immune response.*

I explained how genes can control the types and classes of immune responses and spoke a bit about the classic experiment of Rolf Zinkernagel and Peter Doherty that won the Nobel Prize for describing how T cells recognize their target. "We all have blood types such as A, B, and O," I said. "We also have blood types related to the immune system, which are called *tissue types.* These tissue types are important not only for transplantation, but are needed for T cells to function. Tissue types are governed by molecules called MHC molecules (MHC stands for *M*ajor *H*istocompatibility *C*omplex). Why are MHC molecules so important to T cells? T cells cannot be activated or function without MHC molecules. MHC molecules are on a different cell type than a T cell (called an antigen presenting cell) and the MHC molecules on this cell pick up proteins, process them, and present them to the T cell. The T cell cannot recognize a protein unless it is presented by an MHC molecule. If one imagines that a protein is a hot dog and the MHC molecule is a bun, then the T cell cannot eat (recognize) the hot dog unless it is in a bun. MHC molecules also determine which T cells come out of the thymus which, of course, is crucial for a disease like MS, which is believed to be caused by T cells. To further illustrate how important MHC molecules are, whenever a study is done to look for genes that may be linked to MS, genes for MHC molecules are found. Nonetheless, MHC genes are not the whole story, and we know there are other genes (non-MHC genes) that determine the types of immune responses that occur in MS. In summary, point five says that the class of immune response is related to sets of genes and defining these genes will help us understand why one class of immune response occurs over another and causes MS.

"But genes are not the entire story. Both nature and nurture play a role," I said as I put up the slide with the sixth point.

6. *Environmental factors also determine the class of the immune response to myelin antigens.*

"The environment also influences whether some people develop MS or not," I said, and presented one of my favorite analogies. It is an example that makes a very sophisticated point in a simple way that is easy to understand. It is an example I tailor to the audience I am speaking to, and in this instance I compared Belfast to New York.

"The brain and the immune system both have complex recognition systems," I began. "Both must have memory, and both are dependent on the environment to develop. Indeed input from the environment determines the shape of both the brain and the immune system.

"Let's assume that identical twin girls are born here in Belfast. Because they are genetically identical, one could transplant a kidney from one to the other and it would not be rejected. Indeed, the first kidney transplant was done at the hospital where I work in Boston and was performed between identical twins. Now, to illustrate the effect of the environment on the brain, let's assume that the girls are separated at birth, with one raised in Belfast and the other raised in New York. They both learn language based on what they hear—in other words, based on their environment. We all know that the girl raised in Belfast will speak English with a Northern Ireland accent, whereas the girl raised in New York will speak with a New York accent. The brains of both girls at birth were identical and were receptive to learning language but, depending on the language they heard, the language area of the brain was shaped differently.

"Something similar happens with the immune system. Remember, the immune system is composed of white blood cells that evolved to interact with the environment and protect us against infection. Just as the brain reacts to language that it hears and is shaped in a unique way, the immune system reacts to the infectious agents to which it is exposed in the environment and is shaped in a unique way—it develops its own language, its own repertoire. The immune system of someone who grows up in India is different from the immune system of someone who grows up in the United States.

"For MS there is epidemiological evidence suggesting that infections or exposure to infectious agents in childhood may determine who develops MS. For some people, unfortunately, it may be that in childhood the immune system learns the 'language of MS,' so that later in life, when the immune system is challenged, the body reacts in a way that causes MS. Thus one of the origins of the disease relates to the environment and how it shapes the immune system."

Takashi Yamamura and Takeshi Tabira, MS experts from Japan, told me that in the past Japanese MS was different from MS in Western countries. Japanese MS predominantly affected the eyes and spinal cord. Recently, Japanese MS has begun to resemble Western MS, and there is no evidence that the change was related to genetics. Yamamura postulated that environmental factors were responsible and could include anything from infections to less fish in the diet. Indeed, there are advocates of a diet high in fish and low in animal fat for MS, but the value of such diets is unproven. And despite anecdotal reports, there is no evidence that trauma or stress triggers MS, though smoking does appear to be a risk factor for MS (another reason to stop smoking).

"Some scientists argue that diseases such as MS or juvenile diabetes exist because our environment has changed and is in fact cleaner. Thus the infections one is exposed to are different and make us susceptible to these diseases. The question immediately rises whether we can shape or change the immune system so it does not learn the language of MS. Perhaps this is the way we will ultimately cure MS—by understanding what shapes the immune system and vaccinating or exposing people to the appropriate infectious agents at an early age so that they cannot learn the language of MS. Being exposed to an infectious agent at a certain age can be beneficial. Polio was common in the United States until the introduction of the polio vaccine, but it was relatively rare in Mexico. Why? Children in Mexico were exposed to poliovirus at an early age and became naturally vaccinated.

"There are other environmental factors that affect whether people get MS. MS is more common the farther one is from the equator. No one knows why. It could be there are more viral infections in a cold climate than in a warm climate and these trigger the disease. Some postulate that it relates to the amount of sunlight and exposure to UV light. Sunlight triggers the body to produce vitamin D, which can affect the immune response in a positive way for MS patients by shifting the class of immune response away from Th1 and toward Th2 and Th3. Thus lesser exposure to sunlight the further one is from the equator increases susceptibility. Some argue that we should treat MS by giving vitamin D and that those with a lower vitamin D intake are more susceptible to MS.

"But we know it isn't all environment. There are some people who are genetically resistant to MS no matter what environment—African blacks, Eskimos, American Indians, and certain Asians. I am sorry it is so complex," I said, "but that is the way it is. It is both nature and nurture."

I put up point seven.

7. *MS is not caused by a persistent or single viral infection although infectious agents play a crucial role in the initiation and perpetuation of the disease.*

"For years scientists have searched for the MS virus or another MS infectious agent and still have not found it. If MS were caused by a single virus, it would be so easy. There have been countless searches and many reports of viruses or infectious agents causing MS, but unfortunately these reports have all turned out not to be true. Once we thought MS was related to the measles virus, but after the population was vaccinated against measles, rates of MS didn't change, though another disease related to measles, called SSPE, disappeared. However, as one of my first mentors, Henry Claman, said, absence of proof is not proof of absence, and it is still theoretically possible that MS is caused by an unidentifiable infectious agent, though I don't believe this is the case, and most scientists don't. Nonetheless, as we discussed earlier, viruses and infectious agents are crucial in MS because we believe they are one of the major factors that trigger the immune system to attack the brain. Indeed, there are some animal models where this occurs."

I introduced the next slide by saying, "Point eight is cheating in a way."

> *8. Unexplained defects in regulatory mechanisms and/or tolerance induction exists in MS.*

I translated simply, "Something is wrong with the immune system in people with MS, but we don't know what it is. What makes MS a disease," I emphasized, "is the fact that there are repeated attacks on the nervous system by T cells. If there was only one attack, there would be no disease. The question is, why doesn't the immune system regulate itself? Why can't it shut things off?

"If there was only one attack of inflammation, we would call the disease singular sclerosis rather than multiple sclerosis. The defect in immune regulation in MS is very complex, and it is only through a better understanding of the immune system that we will be able to understand why the immune system doesn't regulate itself. Einstein couldn't understand certain parts of the universe, and in one of his equations he put in an unknown factor called lambda. The unexplained defect in how the immune system regulates itself in MS is my lambda. Possibilities include a defect in one of the types of regulatory cells or in second T cell signals that keep disease-inducing cells under control, something wrong in the thymus that leads to an abnormal repertoire of T cells, or a defect in the body's ability to kill unwanted T cells.

"Nonetheless," I said, "the ability of the immune system to regulate itself in MS patients is not completely gone. There are two stages of the disease: a relapsing-remitting stage, in which people have attacks and recover, and a progressive stage, in which people continue to worsen."

I put up the slide with point nine, a point that also needed no translation.

> *9. Relapsing-remitting MS naturally regulates itself. Treatments which augment these natural regulatory mechanisms will help control the disease process.*

"Understanding the mechanisms that help shut off an attack will help us in finding treatments," I said. "It appears that there are two regulatory mechanisms that are triggered to shut off an attack. First, there is a shift from Th1 cells to Th2 and Th3 cells, and the appearance of other regulatory cells, and second, the Th1 cells that are causing the damage die. That is how the immune system shuts itself off. Remember, after the immune system carries out a function such as fighting off a viral infection it must shut itself down when the infection is over.

I then introduced point ten. "There have been many debates over the years about what causes MS and how to treat it," I said. "We spoke about one of these debates, whether MS is caused by a virus sitting in the brain, and we have declared in point seven that there is no evidence for a viral infection of the brain in MS patients. Another major debate that has occurred since it was first described is whether the animal model we use to study MS called EAE is really representative of the disease. Here's the answer."

> *10. MS is for the most part analogous to the various forms of EAE. Thus there are numerous stages in the immune cascade where the disease can be impacted.*

"The animal model for MS is called EAE, which stands for experimental allergic encephalomyelitis. The history of the study of EAE is in some ways the history of MS. How do we cause EAE in animals? Remember, the very first point in the twenty-one-point hypothesis is that MS is caused by a T cell that is targeted to a protein in the brain and it goes to the brain and initiates damage. That is exactly how one induces EAE in an animal. One takes a protein that is specific for the brain, such as myelin basic protein, mixes it with an adjuvant, and injects it in the animal. By using an adjuvant, one induces Th1 cells instead of Th2 or Th3 cells, and the Th1 cells go to the brain and cause inflammation. One can create animal models where these T cells cause relapses and remissions or models where they cause a chronic disease. In fact, one can clone such brain-specific Th1 cells and transfer only these T cells into an animal and get an MS-like disease.

"With the latest technology researchers have even created genetically altered mice that have increased numbers of T cells that react with the myelin proteins in the brain, and these animals spontaneously get an MS-like disease. Vijay Kuchroo, an experimental pathologist and expert on animal models of MS and autoimmunity in our MS center, recently created a genetically engineered animal in which all the T cells reacted with a protein in the myelin sheath called MOG, myelin oligodendrocyte glycoprotein. He discovered that the animals spontaneously developed optic neuritis, an inflammation of the optic nerve that led to blindness. It was identical to what happened in MS patients. Optic neuritis is often one of the first symptoms of MS. When I asked Vijay how he could tell the animals had difficulty seeing, he told me he discovered it when one of his fellows asked whether it was normal for mice to be scratching at their eyes with their paws. When his article describing the mouse was published in the *Journal of Experimental Medicine* in 2003, the journal put on its cover the original description of EAE by Thomas Rivers, and Larry Steinman of Stanford wrote an editorial in which he described the seventy-five-year history of research on EAE.

"The EAE model is not perfect, and there are different types of EAE. Nonetheless, the immune cascade in EAE provides many targets for therapy, and researchers and pharmaceutical companies are studying this cascade to find new ways to intervene and block the disease. In addition, degenerative changes that occur in MS can also be studied in EAE models.

"Point eleven is another area that was once controversial but for which there is now general agreement," I said as I put up the next slide.

11. MS is driven by T cells that continually migrate into the brain and spinal cord after which local immune reactions may become established.

"Point eleven says that in multiple sclerosis there is a continual movement of T cells from the bloodstream into the brain. Thus MS is not just a brain disease; it is a disease that involves the immune system and can be identified in the blood. This is a very important point, because if MS was localized only to the brain, one couldn't study or access it through the blood. Furthermore, if MS is caused by T cells that move from the blood into the brain, it provides a potential avenue for therapy. Indeed, it has been shown in the animal model of MS that if one blocks the movement of cells into the brain, one can treat the disease. This is now being tested in MS patients using an antibody that blocks a structure on the T cell required for the T cell to enter the brain, and clear, positive effects have been found. However, once T cells get into the brain, they can set up a local immune reaction so that the inflammation in the brain gets a life of its own in addition to cells that may continue to enter the brain. This is the basis for the elevated levels of gamma globulin that we see in the spinal fluid of patients with MS. Clearly, we want to prevent these local immune reactions from becoming established.

"Point twelve is another point that needs no translation," I said, as I put up the next slide.

12. Subtypes of MS exist.

"A major factor that makes MS so difficult," I said, "is that it is not a single uniform disease but one that has different subtypes. The disease can be different depending both on the genetics of the individual and the exposure of the person to the environment. For example, although MS usually begins with a relapsing-remitting course and then becomes progressive, in some patients the disease is progressive from the onset. This is a different subtype of the disease, called primary progressive MS. I have a family I treat in which the mother has the relapsing-remitting form and her daughter has the primary progressive form. Thus, although there was something in the family genetics that caused the MS, it expressed itself differently in mother and daughter. Identical twins can express the disease differently because of different exposure to the environment, which, as we learned, can affect the immune system. Even in the animal model there are subtypes of disease depending on which myelin protein in the brain is being attacked and the strain of the animal.

"There are different subtypes of MS because of the many complex factors that come together to cause MS. Also, there are different ways the immune system can attack the brain, and these different ways can cause different types of MS. How many subtypes of MS are there? Claudia Lucchinetti and Hans Lassmann have pathologically defined four subtypes," I said, "and there may be as many as ten. One of the major challenges of MS research is to understand the basis for the different subtypes so that treatment can be tailored to the subtype.

I turned to the next point. "Point thirteen is a very important point," I said, "as it deals with the concept of what happens when MS changes from the relapsing-remitting form to the progressive form."

> *13. When MS changes from the relapsing-remitting to the chronic progressive form, T cells enter a state of chronic activation and degenerative processes occur.*

"For the MS patient the transition from relapsing-remitting to progressive MS is crucial, because when the disease becomes progressive it leads to the greatest disability and is the hardest to treat. At a recent conference in Venice sponsored by the European Charcot Foundation, I was asked to speak on how to treat progressive MS. Although I outlined the theory behind the treatments being tested, I said that the best treatment is not to let patients become progressive.

"What happens when the patient enters the progressive phase?" I asked. "Two things occur. First, the immune system itself becomes more chronically activated and the immunomodulatory drugs we have don't work as well. Second, the disease changes from being not only an inflammatory process but one that also becomes a degenerative process. We believe that the initial stages of MS are caused by cells that go to the brain and set up recurrent attacks of inflammation. However, over time, factors released by cells set up a degenerative process in which the damage to tissue occurs irrespective of whether there is inflammation. This means treatments directed to the immune system may not be as effective in the progressive phase. The recognition of inflammatory versus degenerative processes in MS has changed our thinking about the disease. It has led to attempts to find different drugs to treat degenerative changes when they occur, and it has made it even more crucial that we treat early to prevent degeneration from occurring. There is evidence that axons, or nerve cells, in the brain are damaged in MS, this begins early in the disease, and when they are damaged they cannot easily be repaired. Thus, the degree to which immune therapy can help the progressive stage remains to be determined."

Before I put up the next slide I asked the following question. "What has been one of the greatest breakthroughs in our understanding and treatment of MS?"

The audience was now primed, ready to answer my questions. Someone called out, "The human genome project." Another said it was one of the new drugs approved for MS.

"No," I said, "it's the MRI scan. Imagine trying to treat tuberculosis without a chest X-ray to monitor the disease or study the heart without a cardiogram," I continued. "We could not study the stars without a telescope or study the world of microbes without a microscope. The same is true for MS. We really were not able to study MS properly or develop a treatment for MS until the MRI scan." I put up the slide.

> *14. Although imperfect, the MRI is linked to the disease process and to the degree to which disability has accumulated.*

"The MRI has allowed us to see MS in a way that we have never seen it before, and although it is not yet perfect, what we see on the scan is linked to the disease process and ultimately to the degree to which disability has accumulated. For example, there is evidence that the more MS there is on the MRI scan, the greater the disability the patient will have fourteen years later. There is also evidence that when there is inflammation on the MRI scan, which appears when dye is injected into the bloodstream and leaks into the brain, patients have more attacks. This fact is used to screen new drugs.

"What are some of the new insights about MS we have learned from the MRI? One of the most important is that there may be progressive loss of brain and spinal cord tissue, a process called atrophy, and in some people this can start at the beginning of the disease and appears to be more closely linked to disability than our previous MRI measures. In addition, although we have traditionally thought that MS only affects the white matter of the brain, with new MRI techniques, we've found that the gray matter is also involved, and there may be disease in parts of the white matter that look normal with conventional techniques. This may be one of the reasons some MS patients have problems with cognition. Thus MS is more widespread in the brain than we thought and this is another reason that early treatment is important.

"We have also been able to see repair processes at work in MS by MRI. We have been able to track new areas of MS by MRI and watch them become smaller, both due to natural repair processes and to MS treatments.

"I believe that future advances in our ability to see the brain in MS patients with the MRI and newer imaging techniques, how it changes with therapy and how it reveals different subtypes—will be one of the keys to ultimately curing the disease. An example of one such new imaging technique is called PET scanning (Positron Emission Tomography), which may allow us to visualize specific cells in the brain. Furthermore, computer programs are now beginning to analyze MRI changes in MS to provide precise quantification." I told the story of the technician that performed by hand the MRI analysis for the Betaseron trial, the first FDA-approved drug for MS, and how her numbers changed and had to be redone when she got a new pair of glasses.

There were only six more points. "When I began studying MS in 1971 there were no treatments for the disease," I said. "That has changed."

I put up point fifteen, which needed no translation.

15. *Many treatments have been shown to help MS.*

The slide contained a list of immune therapies that had been shown to help MS. It was point fifteen and ironically there were fifteen different treatments listed on the slide. I had shown this list to many of my colleagues, and virtually all of them agreed with the treatments that I had listed. The FDA-approved drugs for MS were there: interferon, copolymer-1 (now called glatiramer acetate), and mitoxantrone, along with a number of other immune therapies that had been shown to help MS.

I emphasized that a treatment may have a positive effect on MS even though each trial did not demonstrate the positive effect. Different results could relate to dosing schedules and different patient subgroups being treated. "This is what has caused a great deal of confusion in the MS world," I said. "We know that subtypes of the disease exist and that there are different ways to give medication. Because every patient does not respond to every treatment does not mean that a particular treatment could not help the disease in some patients." I gave the example of antibiotics and the treatment of pneumonia. "Just as there are subtypes of MS, there are subtypes of pneumonia," I said, "some caused by bacteria and some caused by viruses. If we can't tell the difference between the two based on the clinical picture and we treat a viral pneumonia with antibiotics, it won't help. The same is true for a disease such as MS.

"One of the major hypotheses about MS is that it is caused by the immune system," I said, "and the proof of the pudding is that immune therapy would be expected to help the disease. Indeed it has—imperfectly, but it has. The positive effects of many of these immune therapies have been shown not only clinically but also on the MRI scan. The challenge now is to better understand the complexities of the immune system and to find more effective immune therapies that can be given early in the disease.

"What is important conceptually about the list," I continued, "is that all of the treatments affect the immune system in a common way: they decrease Th1 responses, increase Th2 and Th3 responses, or affect movement of cells into the nervous system. Thus we are much farther ahead in MS than in a disease such as Lou Gehrig's disease, in which we still don't have a clear understanding of the disease."

What will this slide look like in the future? First, there will be better drugs to modulate the immune system. Second, a new class of drugs that are designed to affect the degenerative processes in MS and facilitate myelin repair are being developed for MS, and in the future there will be a whole new category of drugs to treat MS on the list.

Points sixteen, seventeen, and eighteen were sophisticated immunological points, and I presented them together. They related to the concept that immune therapies will ultimately be effective by acting on specific T cells that migrate to the brain and are targeted to one of the myelin brain proteins.

16. Treatments that decrease myelin-specific Th1-type cells and increase myelin-specific Th2- and Th3-type cells would be of benefit in MS.

17. Because of the concept of bystander suppression, one does not need to know the exact protein under attack as long as regulatory or protective T cells are targeted to one of the brain proteins.

18. Since MS occurs only in the brain and spinal cord, treatments can work only if they affect T cells that were targeted to the brain and spinal cord.

I put up the slide with point nineteen.

19. Effective treatment will require pulse or continuous therapy and, ultimately, combination therapy.

"Because MS is complex," I said, "until all the factors associated with the disease are known, it is unlikely that a single treatment given once will be effective in all forms of MS. The effective treatment of MS, like the treatment of cancer or AIDS, will require combination therapy, as one drug will not be effective against all parts of the immune system. Furthermore, we also know from the MRI that the disease is more active than the patient realizes. Thus, the disease in some ways is like a sleeping monster, ready to be awakened by environmental factors. Or think of the MS patient as a boxer in the ring. At the present time our treatments can put a shield between the patient and the disease so the patient won't get hit as often or as hard, but we have not yet figured out how to get the patient out of the ring. Because the MS patient is always at risk for another attack, some form of chronic therapy must be given. Now, there can be some exceptions—types of MS that may shut themselves off without chronic treatment, or types of MS that are benign. In these instances we may not need to give continued therapy, but we don't yet know in advance who these people are.

"Point twenty is an easy point," I said, "and a logical extension of developing immune therapy for the disease."

20. Identification of immune measures that are linked to the clinical course will be the cornerstone of therapy in MS.

"At the present time, we have no reliable blood test to characterize the immune status of the patient. If MS is truly an immune-mediated disease, as we believe, then we must be able to measure the overactivity of the immune system and understand when it has been effectively quieted or not. Imagine trying to treat high blood pressure without being able to measure blood pressure," I said. "Thus, along with the MRI scan the identification of immune measures that are linked to the clinical course will be a cornerstone of our treatment of MS." Reinhard Hohlfeld and Roland Martin organized an NIH international workshop devoted to the discovery of such immune markers.

I stopped and took a final drink of water as I put up the last slide. The last slide dealt with the treatment of MS as a complex disease, a theme woven throughout the twenty-one points.

> *21. Because MS is multifactorial and a heterogeneous disease, there will be responders and nonresponders to each effective therapy. The earlier the treatment is initiated, the more likely it is to be effective.*

"I believe a very important treatment concept in MS is that there will be responders and nonresponders to each effective therapy," I said. "Just because someone doesn't respond to a treatment doesn't mean that the treatment will not help others. In fact, one of our challenges is to understand the different subtypes and to develop therapies for individual patients. Finally—and this is a point that one doesn't need to be a rocket scientist to understand—the earlier treatment is initiated, the more likely it is to be effective. A stitch in time saves nine," I said. "It is a general principle for all medicine. Virtually any disease caught in its early stages is easier to treat."

I knew that this was a difficult point to make to an audience of MS patients, as there are always those in the audience whose disease has progressed beyond the initial stages. Thus when I speak about the importance of early treatment, I always point out that research in ways to repair the nervous system is ongoing and that different approaches need to be taken for those who have already suffered damage to the nervous system. Nonetheless, the general principle remains true: the earlier the treatment, the more likely it is to be effective.

"Well, there you have the twenty-one points," I said. "Could any of them be wrong? Of course. If we find a new infectious agent linked to MS, it would change our thinking. The Th1-Th2-Th3 story is not as simple as I've presented it, and there are instances when these cells can have paradoxical effects. There are some scientists who believe the immune response in MS is secondary and isn't really triggering the disease, that MS starts in the brain by an unknown process and is primarily a degenerative disease. And, of course, there are some of the twenty-one points we don't understand completely. Nonetheless, the points represent our current thinking about the immune basis of MS and how it has evolved over the last twenty-five years." When I sent the twenty-one points to Henry Claman, my first immunology mentor who has served as a scientific advisor to the National Multiple Sclerosis Society, he wrote back, "I like the twenty-one points, Howard. Well done."

I took questions, and ended my talk by telling the audience about Normie. When I showed pictures of us together in the assisted living facility where he lived and then as teenagers, I got tears in my eyes. Then I spoke to the large number of people who came up to me after the talk, answering both questions about my talk and personal questions—what drug someone's family member should take, whether I knew a doctor in the United States who could help someone's cousin, whether there was something that could help a person's bladder, and whether I recommended horseback riding or bee stings or a new treatment someone had found on the Internet.

After the hall emptied, Dr. Layward, Jack Pritchard's son (who was himself a doctor), and I had a Guinness in a famous old pub in downtown Belfast across from my hotel. We watched the events in the United States on television, and I fingered the letter opener that had been given to me, inscribed "Pritchard Lecture, Belfast, September 11, 2001." We talked about the twenty-one points and how they were right or wrong. I believed in them, as did most of my fellow scientists, but the ultimate proof would come when a talk about MS could be given explaining MS in historical terms without having people suffering from the disease in the audience.

15

TAMING THE MONSTER

In 1971, when I saw John Saccone with Dave Dawson, there was little we could do for MS patients. More than thirty years later that has changed.

Tuesday morning is my main day for seeing MS patients, although as the number of patients has grown, Tuesdays have spilled over into Wednesday afternoon and other days of the week. I thought back on John Saccone and seeing patients with Dave Dawson as I prepared for my patients on this Tuesday morning. When I was a resident, many MS patients were hospitalized for attacks. MS has now become an outpatient disease, and few patients are hospitalized. I would be seeing patients in our new MS center, which had been in operation since the spring of 2000.

Over the years, the MS clinical service at the hospital had grown, as had the number of doctors and nurses looking after the patients in one of the hospital-based clinics. But the care of MS patients had become fragmented—infusions were given in the chemotherapy suite in the hospital, and we struggled to find time enough on the hospital MRI machine for our patients. As our understanding of the complexity of MS increased, the number of treatments available increased, and the importance of MRI imaging became clear, it also became clear that we needed a separate MS center. I was sitting with Ferenc Jolesz, head of MRI imaging at the hospital, discussing ways to study MS patients and follow them clinically by MRI when he said to me, "Howard, the only real way is to get your own MRI machine and your own image processing laboratory."

"How much does the MRI cost?" I asked.

"A million and a half dollars," he said, "with a discount."

I told him I was not afraid of the number and set out to raise the money.

The path was tortuous. It not only involved raising the money, it involved hospital politics, radiology politics, and finding the space. Nonetheless, we succeeded, though the amount needed grew from $1.5 million to $2 million and then to $4 million. Although the crisis in medicine that affected hospital finances was causing a crisis in medicine for us, and for our plans for building an MS center, there was a silver lining. Because of financial pressures, the Brigham and Women's Hospital and Massachusetts General Hospital combined into one entity called the Partners HealthCare System, which meant that the long-time cross-town rivals were now working together. This allowed us to create one place where all MS patients from both hospitals could be seen. The MS center had four components: a clinical center where patients were evaluated by the doctors, an infusion unit where patients received IV therapy, a dedicated MRI machine with an image processing laboratory, and a clinical immunology laboratory adjacent to a basic immunology research laboratory to investigate and measure the immune system in MS patients. The MS center has larger-than-normal exam rooms, lower exam tables with hydraulic lifts, and a red-carpeted area for the twenty-five-foot walk that is part of every patient evaluation. The MS center's MRI and image processing laboratory were only two blocks away, headed by Charles Guttmann, and MRI images could be shown to patients via a computer hookup at the MS center. I remember early one Saturday morning sitting on a crane with Charles that blocked Longwood Avenue as it moved the MRI magnet from the back of a truck through a hole in a basement wall just below my old research office.

I felt a great deal of satisfaction as I put on my white coat and prepared to see MS patients on this particular Tuesday morning: the satisfaction of having watched the field come so far in thirty years, and the satisfaction of setting up the MS center. As it turned out, Dave Dawson was also part of the MS center and we often discussed difficult MS cases together, with options of therapy and nuance of diagnosis that didn't exist when I presented the case of John Saccone to him in 1971. Nonetheless, underneath the satisfaction there was frustration and even anxiety, because I knew there was still a long way to go.

Susan Gauthier was a neurologist interested in clinical studies of MS and would spend the day with me seeing MS patients. She had finished her neurology residency and was spending three years as a clinical fellow studying MS. MS clinical fellowships didn't exist until recently. With five FDA-approved drugs for MS and countless others waiting to be tested, it became clear that the training of clinical neurologists to carry out clinical trials and clinical studies in MS was needed. Such fellowships were created by the National Multiple Sclerosis Society and NIH. Susan's interest was in MRI imaging and clinical trials, and she would also take courses on statistics and epidemiology at the Harvard School of Public Health. She became interested in MS when one of her classmates came down with the disease, and she had spent time with Larry Jacobs at Buffalo.

The first patient we saw was a new patient who had never been seen before at our MS center. She was twenty-seven years old and came with her husband and her mother. I escorted her into the exam room, and Susan joined us. She brought with her a series of MRI scans and medical records from her primary care doctor and neurologist. She was a tall, thin, attractive woman, with straight brown hair that rested on her shoulders. She wore glasses, which gave her a studious look, and indeed she was a teacher. Her husband worked as an engineer at a computer company. I looked at the form that she filled out and noticed that she had one child. In the area of the form where medications were listed, she had written "none." In response to questions related to MS history, she had noted dizziness as a symptom.

"How can I help you?" I asked, as I slid forward in my chair and placed her MRI scan on the counter.

"My doctor thinks I might have MS, and I have come for a second opinion," she said.

"A number of medications were recommended to us but we didn't want to start anything until we were sure of the diagnosis and spoke to someone such as yourself," the husband said.

I briefly looked through the records she had brought, and read the doctor's last note. Although the details of her case were in the doctor's note, it was always better to get the history directly from the patient, to gather primary information.

"I notice you have one child," I said glancing at the intake sheet. "Is it a boy or a girl?"

"A girl," she said.

"How old is she?" I asked.

"She just turned three."

"What's her name?" I asked.

The woman smiled.

Asking someone about their children puts them at ease and a mother always smiles when she says her child's name. Children are important in MS because most MS patients are of childbearing age, which can affect the use of medication and invariably there are questions about having children and its effect on the disease. Also there is an increased tendency to have MS attacks after the birth of a child. I asked her about other medical illnesses or problems. There were none. Many patients were young and healthy apart from the MS.

"Tell me about the symptoms that led to your MRI and question about MS," I said.

"My problem began about six weeks ago," she said. "I was playing with Jennifer on Saturday afternoon and began to feel dizzy. I didn't think much about it, but the dizziness got so much worse that I had to spend most of Sunday in bed. When I couldn't go to work on Monday, I went to my doctor. The doctor gave me some pills for dizziness, meclizine, but they didn't help. I was out of work for the entire week, until finally my doctor sent me to a neurologist. It was a neurologist who ordered the MRI scan."

I looked at the doctor's note and the neurologic exam, which was normal, save for some nystagmus or jerky eye movements.

"The doctor called me the following week and told me that the MRI suggested I might have multiple sclerosis."

"Did you get any treatment, any steroids?" I asked.

"No, I didn't want to take steroids, and I was beginning to feel better without treatment."

"Have you ever had any symptoms like this before?" I asked. "Ever lose vision in an eye? Have weakness in an arm? Numbness?"

"Something happened a couple of months after Jennifer was born," her husband said, "though we're not sure it meant anything."

"What was it?" I asked.

"She developed a funny feeling in her legs. She described it as if she was walking on sand. She had been to the beach a day or two before, and I actually wondered if she had burned her feet on the sand."

"Was the numbness just in her feet, or did it involve her legs as well?"

"It went up the legs as well, and I even felt numbness in my chest," she said, drawing a line with her hands just below her breasts.

"Did you see a doctor or have any tests done?" I asked.

"No, it got better on its own and never happened again."

As I made notes in the chart, I was fairly certain that her symptoms represented the first attack of MS. It was classic to have symptoms such as she described, especially numbness up to the waist. Anatomically it was probably in an area affecting the spinal cord. Furthermore, the fact that it had come on two months after the birth of her first child was consistent with the known increase in MS attacks that occurred after pregnancy.

"Time for your examination," I said.

First we stepped into the hallway and I watched her walk twenty-five feet on the red carpet in our MS center. She did it in four and a half seconds, and her gait was normal. Back in the exam room, I performed a detailed, careful neurological exam. I couldn't detect any abnormalities in her eyes, strength, sensation, or reflexes.

"Your exam is excellent," I said. "Dr. Gauthier and I are now going to view your MRI scan, and then we'll be back," I said, as I picked up the films and left the exam room.

When we put up the films on the view box in our viewing room, it was clear that her MRI was abnormal. There were films of her head and of her thoracic and cervical spine. The head MRI showed two areas of abnormality near the ventricles and one in the brain stem. It was probably the abnormality on the MRI in the brain stem that led to her dizziness. There was also one area of abnormality in the thoracic spine that may have accounted for the numbness she reported after the birth of her first child. We then looked at films taken after the injection of gadolinium. Gadolinium is a contrast agent injected into the vein that shows areas of active inflammation in the brain and breakdown of the blood-brain barrier. There was some gadolinium enhancement in the brain stem area, but nowhere else.

"It certainly looks like MS," I said to Susan. "Two attacks and areas of abnormality on the MRI both in the brain and the spinal cord. There is now a new classification of MS based on MRI imaging. In the past one needed two attacks separated in time to make a diagnosis. Now a diagnosis can be made after one attack, and new disease activity on the MRI scan can qualify as a second attack."

I took down the films and we went back into the exam room.

"I don't think there is any question that you have MS," I began as the patient took out a notebook and began writing things down. I noticed the notebook also had a long list of questions. I explained that the first attack had most probably occurred three years earlier, after the birth of her child, and the dizziness was a second attack. The MRI was consistent with the diagnosis of the relapsing-remitting form of the disease.

"Do you recommend that my daughter begin therapy now?" her mother asked.

"Yes. We believe the earlier one begins treating a person with MS, the better the chance to help. When I began working on MS more than thirty years ago," I said, "we didn't have any therapy to offer. Fortunately, that has changed."

"But treatments are given by injections," the husband said, "and my wife hates needles."

Many patients hate taking injections. "At this time, all the approved drugs for the relapsing-remitting form of MS are given by injection," I said. "However, we have research trials here at our center in which we are testing new oral therapies, and you can participate in the trials if you like," I added.

"But isn't it true that the experimental therapies have not yet been proven to show a beneficial effect in MS?" she asked.

"That is correct," I told her.

"I think you should take a proven medication," her mother said.

"We read on the Internet about the different types of injectable therapies," her husband said. "Which one do you recommend?"

"I don't recommend one over the other," I said. "It is important for you to understand what is involved with each of them and then come to a decision about which to begin with. What is most important is that you begin some type of therapy. If you can't tolerate one therapy or want to switch, you always can."

I then went over the four drugs. First the interferons: Avonex, which is given weekly by an intramuscular injection; Betaseron, which is given subcutaneously every other day; and Rebif, which is given three times a week by subcutaneous injection. I explained that the interferons had side effects, including flulike reactions and, potentially, injection site reactions, but that the side effects usually lessened with time and could be helped by taking Tylenol, although as with any medication, the degree of side effects depended on the patient. I explained that I had patients who have been taking the interferons for years with no problems and others who found them difficult to take. I then explained Copaxone, which was a daily subcutaneous injection and in general had less side effects that the interferons.

"What would you take if you were me?" she asked.

As I prepared to answer, I imagined that there was a secret video camera in the room with representatives of each of the drug companies watching to hear what I said and what I would recommend to my patients. In a world in which MS drugs cost over $10,000 per year per patient, companies want doctors to prescribe their medicine. That's why they would love to have those secret video cameras. Even without the video cameras, drug companies can gain access to the prescribing patterns of every doctor in the country that prescribe their drug, and doctors who tend to prescribe one company's drug are often invited on medical educational junkets set up by that company. This is true for other diseases as well. Indeed, doctors at different MS centers or in practice are known to be Avonex doctors, Betaseron doctors, Copaxone doctors, or

Rebif doctors. In general, doctors either let a patient decide after giving the patient four boxes prepared by the drug companies, each of which has a pamphlet and videotape that explains each of the injectable drugs, or the doctor makes a recommendation after attempting to understand the patient and what would be most appropriate for him or her.

In my mind, more important than which drug is initially prescribed is that patients be closely monitored over time and treated for any new activity that develops, since the drugs are only partially effective in most patients. I gave this patient the materials and the videotapes that explained each of the drugs and suggested that she read through them and then either come to a decision or give me a call with further questions to help with the decision. It was often better for the patient and family to confront these issues at home, away from the doctor's office, after a good night's sleep.

"What about having more children?" she asked.

"There is no reason that you can't have more children," I said. "However, you will have to discontinue taking the medications prior to getting pregnant."

"I read that when a woman is pregnant the MS actually gets better," the husband said.

"That's correct," I said.

"Why is that?" the mother asked.

"The immune system is suppressed during pregnancy, and that helps MS, in which the immune system is overactive."

"Well, maybe I should get pregnant right away," the patient said, "and by the time we have the next baby there will be oral medications." Indeed, there are strategies to develop drugs based on factors associated with pregnancy. Rhonda Voskuhl found in an initial pilot trial that oral estriol, a hormone that increases during pregnancy and is associated with a Th1 to Th2 shift in the immune system, reduces attacks and benefits the MRI in relapsing-remitting MS patients. In addition, alpha-feta protein and interferon tau are pregnancy-associated factors that have shown positive results in the EAE animal model and will be tested in MS.

I laughed, remembering one patient who had asked whether she could treat her MS by continually staying pregnant. "I don't think we will have oral medications that quickly," I said, "although I am sure that one day that we will be getting away from injections. My advice is to think about wanting to have another child versus starting the medication now. Your exam is excellent, and the MRI shows only mild changes."

"What if I have an attack after the next child?" she asked.

"Don't worry about it," I said. "Having a family is what is most important, and we can monitor you closely and treat you should you have an attack. Remember, most people do not have problems after pregnancy."

She then read through her long list of questions, addressing diet, exercise, stress, and alternative forms of therapy. They were common questions, and I answered them as best I could, though many did not have the black and white answers that many patients wanted. She agreed to give a blood sample for our immunology research and we enrolled her in our natural history study.

I finished by saying, "Our understanding of MS and our ability to treat it has advanced significantly in recent years, so I think that there is a good chance that we will be able to keep you healthy so you can raise your family and live a normal life."

"That's good news," the mother said.

I PICKED UP the next chart and read through the remarkable case. Jerry Devine, a college professor from Connecticut. I thought back to when I had first seen him and his wife, a year and a half earlier. Jerry was fifty-two and had been diagnosed with multiple sclerosis six years earlier, when he had numbness and double vision. He came to the MS center because of marked worsening of his MS that had occurred over a six-month period. I remembered taking him into one of our larger examination rooms because he was being pushed by his wife in a wheelchair.

"How long have you been using the wheelchair?" I asked.

"Just this last week," he said.

"Three weeks ago he could use a walker," his wife added.

"Three months ago I was using a cane," he said.

"Six months ago we were in London and he was walking normally," his wife said.

"A year ago I could run four miles a day," he said.

"Have you been on any medications?" I asked. He had been treated with injectable medication and switched from one to the other when he began having trouble with his walking, but there was no benefit. He also had received three courses of IV steroids with temporary benefit but no effect on the progression of illness.

"Let's take you out into the hallway," I said, "and see how you can walk."

"I can only take a few steps," he said.

"Do the best you can," I answered as we stopped his wheelchair at the beginning of the red carpet. He was able to stand but was only able to take a few steps holding onto his wife.

"Six months ago you were walking normally in London?" I asked again, just to be sure.

"We walked on the streets, we walked in museums, we walked to the theater."

I wheeled him back to the examining room and performed a neurological exam. There was decreased sensation in the legs, he was unable to move his entire left side, and the big toes on both feet responded abnormally (went up), indicating damage to the motor system.

"When was your last MRI?" I asked, and found out that it had been done eight months earlier.

I left the exam room and reviewed the MRI. The MRI had been done before his marked deterioration, but there were multiple areas of involvement, and one area was enhanced when dye was injected. I showed the MRI to Dave Dawson and Lynn Stazzone, our nurse practitioner.

"He's crashing," Dave Dawson said. "He has not responded to the injectable medicines or to steroids."

"He needs chemotherapy with cyclophosphamide," Lynn Stazzone said.

I agreed. "We can give him monthly treatments, or we can put him in the hospital and give him intensive chemotherapy," I added.

"I think he needs to be hospitalized," Dave Dawson said. "I don't think we can wait a few months to see whether there is a response. But we should get an MRI before we start, just to make sure we are not missing anything."

I called our MRI center, and we sent him over for an emergency scan. The scan showed even more areas of involvement, and now several areas were enhanced when dye was injected. The brain was inflamed in many places, and it was affecting critical areas in his brain stem and cerebellum. I sat with Jerry and his wife and explained the situation.

"I think we need to put you in the hospital and treat you with chemotherapy," I said. "You have a very aggressive form of MS that has not responded to therapy, and if nothing is done you may never be able to walk again. We could give you the chemotherapy on a month-by-month basis, but I don't think we have that much time."

"What does the chemotherapy involve and how dangerous is it?" he asked.

I explained the use of cyclophosphamide and thought back to 1983 when we published our article in the *New England Journal of Medicine.*

"You will be in the hospital for eight days, and you will get five IV treatments with the cyclophosphamide plus treatment with IV steroids. You will have to drink a lot of fluids, and you will lose your hair, although all the hair will come back. Apart from nausea and vomiting, which usually can be controlled with medications, the major risks are potential damage to your bladder and a risk of infection when your white blood count goes down. There is also a theoretical long-term risk that you may be at increased risk of developing other cancers." I gave him a form that explained all the side effects.

"How common are the bad side effects?" he asked. "The risk of other cancers and bladder damage?"

"They are not that common," I said, and assured him that we had treated many patients with this regimen. I didn't want to scare him so much that he wouldn't take the treatment, but I also didn't want to tell him that there were no risks. Nonetheless, in my own mind, I was fairly certain that the treatment would help, given that he had a rapidly progressive course and the MRI showed so much inflammation.

"What are the chances that this will help?" he asked.

"What do you mean by help?" I asked.

"Stop the progression," he said.

"Make him able to walk without a wheelchair," his wife added.

I didn't know what to say. In my own mind, given the type of MS he had, I was confident that there was a 90 percent chance that this would help him. Could I be that bold and tell him that? What if I was wrong? Was it a fair thing to say?

"The chances are fifty-fifty," I said.

"What are my other options?" he asked.

I discussed other potential therapies, including the use of plasma exchange, the use of intravenous immunoglobulin, and the use of other chemotherapy drugs for MS such as mitoxantrone. But in a situation such as his, we'd had the most experience and success with cyclophosphamide, and I thought it was the drug of choice.

"What would you do if you were me?" he asked.

"I would take the treatment," I said. "I think it has a good chance of helping you."

"If he were your brother, what would you recommend?" the wife asked.

I thought for a second. "If he were my brother, I would recommend the treatment."

He looked at me for a second and then smiled. "Do you like your brother?" he asked.

We all laughed.

I put my hand on his shoulder. "You'll be okay," I said. "We're lucky to have a treatment that I honestly think has a good chance of helping you.

Eighteen months later I saw Jerry for his third six-month checkup after receiving the treatment.

"Let's see how you do on the twenty-five-foot carpet," I said.

Jerry stepped onto the carpet and walked the twenty-five-feet in eight seconds without using a cane, although his gait was wobbly. As before, his neurological exam showed decreased sensation in the leg and upgoing toes. An MRI scan done six months earlier showed a disappearance of all the inflammation in the brain. Jerry had begun to improve after he left the hospital and was placed on monthly outpatient boosters of cyclophosphamide plus methylprednisolone for a year, and now was taking them every six weeks. At times there was nausea and a feeling of being washed out after the treatment, but he was able to drive and work full time at the college.

"My memory is not as good as it was before, and there are still problems with my bladder," he said.

"Still, it's a miracle," said his wife.

"I wouldn't go so far as that," I said, "but it is wonderful to see how well you've done."

"What is the plan?" he asked.

"We will continue you on the IV booster therapy which you will receive every six weeks for a year, then every two months until the third year." He was receiving his treatment in the infusion unit at our MS center, and his white blood count was being followed carefully by the nurse practitioner, Lynn Stazzone. "At some point we will start you back on injectable therapy to keep you stable. We can't give you the chemotherapy forever."

"I remember you telling me you would give the treatment to your brother," his wife said.

"And that you liked your brother," he added with a smile.

"I must confess," I said, "I don't have a brother, but I would give the treatment to my sister and I love my sister very much."

I turned to Susan after the patient and his wife left to make the next appointment. "This is an extreme case," I said. "Rapidly progressive disease, marked inflammation on the MRI. We generally don't use the one-week hospital induction regimen with cyclophosphamide anymore—we usually begin with outpatient monthly infusions—but I think it was important for him to be treated quickly and aggressively. We could have given him chemotherapy with mitoxantrone, but I don't think the onset of action would have been fast enough. There has also been recent interest in plasma exchange for people who are in the midst of a severe attack and haven't recovered. What his case shows is that in some instances MS can be extremely aggressive with a strong inflammatory component, and we have a way to stop it."

"What will happen to him?" Susan asked.

"I hope he will remain stable after we put him on one of the injectable therapies," I said.

THE NEXT PATIENT was a young woman who was visiting our MS center for the first time. She was accompanied by her boyfriend and brought with her a large folder with MRI scans and a manila envelope with many pages of doctor's records. On her intake form she wrote that her MS was getting worse and she had difficulty taking the injectable medications.

"Nothing seems to help," her boyfriend said, "and she is in a great deal of pain."

"How did your MS start?" I asked.

"Four years ago I began having pain down the side of my neck. I didn't know what caused it," she said. "They did X-rays of my spine but saw nothing. Then I developed numbness in my hands and had difficulty writing. I went to a doctor, and he found nothing on exam. The next couple of years were difficult for me. I was trying to finish up school and was working at the same time. Also my mother was very sick and I was having to help out with her."

"That's when I met her," the boyfriend added. "She was really under a lot of stress."

She took his hand and they held onto each other as she told the rest of her story. "I then had trouble with my vision—I saw spots in front of my eyes—and had dizziness. My cousin has MS and I have a girlfriend with MS, and I noticed that my symptoms were similar to theirs. That's when I developed problems with my bladder and couldn't go to school anymore because of tiredness and pains in my legs. I finally went to a neurologist and asked that a MRI scan be done. That is when they found the MS."

"Did you have a spinal tap?" I asked.

"No. I was afraid to have that done because of the headache."

"We then started reading a lot about MS and the different treatments," her boyfriend said.

"They told me it was important I get treated as early as possible. So I went on one of the interferon drugs. I couldn't stand it. I was on it for three months and had terrible reactions."

"What were your reactions?" I asked.

"I had all the flulike symptoms, and I was tired. It was horrible."

"That's when they switched her to Copaxone," her boyfriend said.

"I couldn't take the Copaxone either," she said. "I had reactions when I got the injections, and even though you weren't supposed to have symptoms from the Copaxone, it made me tired."

"Did you ever get intravenous steroids?" I asked.

"I got them twice," she said. "They helped with the pain for two or three days, but it didn't last."

"We are trying new types of therapy," the boyfriend said. "Acupuncture and herbal treatment."

"What do you think about those things?" she asked, giving me a list of additional MS treatments she had printed from the Internet. "You think they could help me?"

I carefully looked at the list and then through the records—the notes of the doctors, all the blood tests and all the X-rays and MRIs that had been done.

"Let's go through the neurological exam first," I told her.

We stepped outside into the hall and I had her walk the twenty-five-foot red carpet. She couldn't walk it alone and held onto her boyfriend. Her gait was tentative and at times she almost fell into the wall. Back in the exam room, I performed the neurological exam. She had weakness of her right side when I tried to push her leg down and sudden withdrawal of her toes when I tested for abnormalities of the motor system by stroking the bottom of her foot with a key to see if her toe went up.

"Dr. Gauthier and I want to look at your MRI films, and then I will be back," I said as I headed to the viewing room.

We put up the films of the brain and of the spinal cord and went over them carefully. I noticed two tiny spots in the white matter of the brain. There was nothing in the spinal cord.

"What do you think?" I asked Susan.

"I don't know," she said. "It's a tough case. She doesn't have much on the MRI and she hasn't been able to tolerate injectable therapy. I don't know what to do next."

"Do you know what I think?" I said. "I don't think she has MS. Although her symptoms could be related to MS, they are not classic. Furthermore, I think her problems walking are not real—look how she lurches from side to side. There are also inconsistencies in her exam: the weakness in her legs is intermittent, there are no hard neurological signs, and the MRI scan is not clearly abnormal and would not satisfy current criteria for MS." Now that we have the MRI we frequently are sent cases of people who have been given a diagnosis of MS based on nonspecific changes on the MRI.

"What do you think she has, then?" Susan asked.

"I don't know," I said after thinking a bit. "But I see no evidence for

MS. I am not surprised she couldn't tolerate injectable medicines. There must be something else going on in her life, and I think she is searching for help."

"What are you going to tell her?"

"It's tough. I think it's important that she see a psychologist and I don't want to take away from her the possibility that she has MS, but I think she has to realize that most likely something else is going on. If she does not have MS, we are not going to help her by pretending she does and giving her MS medicine. Anyway, she can't even tolerate the medicines. She is going to need support from her family and other doctors."

We went back into her room, and I faced her and her boyfriend. Before I could speak she pulled out a list of questions.

"I have some questions I'd like to ask you," she said. "First, do you think I need chemotherapy? I heard that there are new chemotherapy drugs used for MS, and maybe that would help me."

She then read through a series of questions about viruses, about diets, about other diseases, and about three other therapies for MS that were currently in clinical trials

"Heather," I said, "before going any further, let me tell you what my thoughts are."

I explained to her that I didn't think she had MS, that there might be other things causing her problem.

It was very difficult for her, and she kept challenging me. "Why do I have the pain? Why do I have problems with my bladder? Why are there changes on my MRI scan? Why do the other doctors think I have MS?"

I explained to her that there might be something wrong with her nervous system, but that I didn't think it was MS and that she was lucky that there was no damage that we could detect. I asked her if she wanted to get better.

"Yes, of course," she said.

I told her the only way to get better was to try a different approach to her symptoms. I emphasized that perhaps this was why the MS medicines she tried hadn't helped her.

She said nothing. Her boyfriend was a bit more receptive. "We have to get you well," he said. "We can't go on like this."

I told her that the next step was for me to talk to her physician and for her to see the psychologist at our MS center. I emphasized that we wanted to get her better and that this was the way to do it.

"Is there still hope?" she asked.

"Of course there is," I answered. "But you must be ready to address your symptoms in a different way and get help and support from a different series of doctors."

"THE NEXT PATIENT is much different from the last," I said to Susan as I read through the next chart. "He's a doctor."

Jeffrey Micola was a family practitioner from Rhode Island whom I had been following for almost seven years. His symptoms began typically, with numbness in the legs and an attack of optic neuritis. His diagnosis had been made by MRI scan, and when the interferons became available he began treatment. As I escorted David into the room I watched his walk, which was good but slightly off.

"This is Dr. Jeffrey Micola," I said to Susan, and he extended his hand. I felt it was important to introduce him as a doctor, both out of respect and because our discussion would be different from the conversation we'd have with someone who wasn't a doctor. I remember when I had elective surgery to repair an anterior cruciate ligament that had torn when I was playing basketball with my sons, and being asked by nurses and anesthesiologists if I wanted to be called Howard or Dr. Weiner. My immediate response was Howard, as I wanted to be friendly, but later I changed my mind and said I wanted to be called Dr. Weiner. I realized that the staff needed to be aware that I knew about medicine and they could use a different language in talking to me. I have found that talking to doctors who are MS patients is different because there is nuance to the dialogue and doctors are always trying to make their own diagnosis and medical evaluation. There are several doctors with MS whom I treat. Two women doctors I treat found support for each other in our infusion center—a chance to talk to someone in the same situation, share stories, and speak about the disease as physicians.

It was easier for me to empathize with a patient who was my age or who reminded me of a family member. John Saccone, one of the first MS patients I treated, had two children, like me. Then of course there was Normie. But doctors with MS were the easiest to empathize with and their cases the most threatening—I couldn't help but think, "There but for the grace of God go I" every time I spoke to a doctor with multiple sclerosis.

"Here is my latest report," Jeffrey said as he handed me three typewritten sheets.

It is not uncommon for doctors to bring in detailed histories, and Jeffrey had prepared three sheets of paper with his latest list of medications, blood tests, MRI results, and a graph he had made showing when he was on which medications, when he had treatments, and the course of his illness.

"You don't leave any work for me," I said with a smile.

"What I do for my patients I might as well do for myself," he said. Jeffrey was a general practitioner who also treated a number of patients with multiple sclerosis and often called me for advice about their cases. During the initial stages of his disease, he had kept his illness confidential, but now as the disease began to take its toll, he told his MS patients that he too had MS.

"How is work?" I asked.

"Good. I get a little tired but I am still able to carry out a full workload."

I looked at the drugs he was taking. He was on an injectable interferon and medication for his bladder, and when he had worsening, he had taken short courses of IV steroids.

"You think it is time to add something more to the treatment program?" he asked, almost reading my mind.

"Let's do the exam and check the MRI first," I said.

His exam was not that bad. He walked the twenty-five-foot red carpet in seven seconds with only a hint of a problem with his gait. The only abnormality on the exam was an upgoing toe on the right and some minor coordination problems. But there was one thing that was bothersome: the MRI scan. We had ordered an MRI scan a few days prior to his visit so we could review it in the context of his clinical situation. Susan joined us, and the three of us sat down at the computer in our viewing room in the MS center and brought up the images. With my finger on the mouse I was able to move up and down his brain, slicing it in different ways and checking for areas of involvement that were enhanced after injecting gadolinium—these were areas with active inflammation. We put up the scan that he had had six months earlier on one screen and compared it to the MRI he had just had done. There were three areas that were active and lit up with gadolinium. In addition, there was one new lesion or spot that hadn't been seen before. The changes weren't dramatic, but it was clear that his disease was not under complete control.

"What do we do now?" he said.

"How are you feeling?" I asked.

"I am feeling fine," he said. "I haven't really slowed down, and although I know there is new activity on the MRI scan, I haven't changed that much and I don't feel it."

"How about your walking?"

"Walking is fine, although I used to be able to walk around the lake near our house and it is now a bit more difficult. I am also a little more fatigued and I am taking naps more."

"How many hours a week do you work?"

"Sixty hours with nights and weekend calls every three weeks."

"Maybe you should slow down," I said, with a smile. "Maybe that's why you're tired."

"Work is what keeps me going. I love being a doctor, and I am not going to give in to this disease," he said emphatically.

When we discussed a case such as his at our Wednesday clinical conference, Samia Khoury, who codirects the MS center with me said, "I don't think any activity is acceptable. We don't want one attack. We don't want one new lesion on the MRI. If that happens, we need to discuss if there are ways to stop it further."

"Should we try shutting things down even more?" I asked Jeffrey.

"I have been thinking about that," he said. "Right now I am taking interferon plus intermittent steroids. I am taking as much interferon as I can, so we can't increase the dose anymore."

"What do you want to do?" I asked. I had my own thoughts but wanted to give him a chance, as I knew he had been thinking about it.

"Do you think it is time to go on one of the chemotherapy drugs?" he asked.

"I think so," I said. I had spoken to him about it before and had just come back from a meeting in New York where the National Multiple

Sclerosis Society was trying to prepare guidelines for what physicians should do when there is an incomplete response to therapy in MS patients.

We had a long discussion about his taking mitoxantrone, which would be given by infusion at three-month intervals or cyclophosphamide, which he would take monthly.

"How about my heart?" he asked.

"We will need to check that if you take the mitoxantrone," I told him. "And you can only take the drug for two years."

"I understand," he said, "but I think it best to keep things under control now, and we'll cross the next bridge when we come to it."

We shook hands, and then he asked advice on the treatment of two MS cases he had in his practice.

"DO YOU LIKE perfume?" I asked Susan as we prepared to see the next patient.

"What do you mean by that?" she asked.

"It will become obvious in a few minutes," I said.

The next person we saw was a fifty-three-year-old woman who was accompanied by her husband. I had been following her for almost twelve years and she came all the way from California to see me. I introduced her to Susan, and we prepared to enter the exam room.

"Do you want to see her walk first?" her husband asked.

"Might as well," I said, and put her at the beginning of the red carpet. She walked down it with a smooth gait in five seconds.

"How does the walking feel?" I asked.

"Feels fine," she said. "I remember when you tested my walking in the old hospital clinic and would measure the distance by counting tiles on the ceiling."

"Yes," I said, "but we've rolled out the red carpet for you here."

I leafed through her chart. "I saw you a year ago," I said as I reviewed the last exam.

"The time really seems to go by," she said.

"Any problems over the last year?" I asked.

"Nothing major. Occasionally I have tingling in the fingers on the left side and every once in a while when I put my head down I feel tingling down my neck."

"Lhermitte's sign," I said to Susan.

As I looked through the notes I noticed she had had a Lhermitte's sign off and on for many years. Other symptoms included dizzy spells that lasted approximately five minutes and mild fatigue and achiness in her leg that happened off and on. Her first symptom of MS was in 1978, almost twenty-five years ago. Her MRI showed lesions consistent with MS, but no enhancing lesions and no black holes, which signify damage to nerve cell bodies. I went through the exam in detail. Normal.

"She's doing well," her husband said, "although at times she worries about the tingling."

We had obtained an MRI the week before, and it showed no new disease activity.

"What about the injectable drugs for MS?" she asked. "I keep reading about how important it is to take them and to start them early, but I have done so well for so many years. Do I really need to take them?" she asked.

"That's a very good question," I said. "The general answer is yes. We believe that all MS patients should be on medicine and it should be started as early as possible. However, we know that there is a small group of patients who do extremely well with their MS without treatment. We call that benign MS. The only problem is that when someone comes down with MS, we can't tell who will be benign."

"I have had MS for so long and I am doing so well," she said. "I think I have benign MS."

"She exercises at the gym and goes for long walks," her husband added.

"The MRI is good, too," I told them.

Later I had a detailed discussion with Susan about the concept of benign MS. "Is there such an entity?" I asked rhetorically. There had been a number of investigations on benign MS, defined as MS that does not lead to disability over at least fifteen years. When I gave my twenty-one-point talk in Belfast, I spoke to one of the doctors there who had studied patients who had been stable for fifteen years. Unfortunately, when they continued to follow these patients some of them began to worsen. Nonetheless, it was clear that there was a small group of patients whose MS was not as severe. It was a fertile area of research. What was different about these people?

It was hard to tell a patient who had done well for twenty-five years that she had to go on therapy, especially if she was resistant to taking injections. With the MRI scan we could follow her closely, and if there were symptoms we could initiate therapy. I wondered if there would be a question in her mind about taking therapy if there was an oral therapy that didn't involve injections.

"Am I making a mistake by not taking treatment?" she asked.

"I don't think so," I said, "as long as we follow you and obtain MRI scans. Are you afraid of what might happen to you?" I asked. "Do you want to take treatment?" I thought in my mind of many patients at the beginning of their illness who were anxious to begin therapy.

"I would rather not be on one of the injectable treatments," she said finally. "I have done so well, and for whatever reason I think my body is controlling the disease."

I couldn't argue with her logic. "That's fine," I said, "but I want to watch you carefully with yearly MRI scans, and if you begin to have symptoms, we will initiate therapy."

Her husband then pulled out a small plastic bag from the satchel he was carrying.

"Here are the latest perfumes," he said as he handed the small plastic bag to me. Her husband was a perfume designer, and on each visit he brought me new representative scents in bottles with a handwritten number on them. I gave some bottles to my wife (after I approved of them, of course) and some to personnel at the MS center.

"Two are for women and one is for men," he said.

"Thank you very much," I said with a smile. "I am beginning to get a collection of these, but remember I don't wear perfume."

"Give it to one of your male friends," he said. "Actually there are some women who prefer perfume designed for men."

"I'll see you in a year's time," she said, "and will get an MRI just before then."

"Call if there are any problems," I said, after which Susan and I smelled the perfumes.

I PICKED UP the next chart and leafed through it with Susan. I had first seen this woman ten years earlier, when she was twenty-eight and had early symptoms of MS that she had noticed while playing tennis. She had been a very good singles tennis player and found that her leg was cramping up on her and she couldn't move around as quickly. She gave up singles and moved to playing doubles. We gave her a course of IV steroids without much benefit.

"Is there anything that can help me?" she had asked me ten years ago.

"I am not sure," I said.

"What type of MS do I have?" she asked.

"Yours appears to be a primary progressive form of the disease," I said. I reviewed her history again. There was no evidence of attacks, no optic neuritis, no numbness, no double vision, no clumsiness.

"Are there any treatments that have been proven to work in the primary progressive form of the disease?" she had asked.

Unfortunately, I knew the answer was that there were no proven treatments. There was only marginal evidence that interferons could help primary progressive MS. A recent trial of Copaxone in primary progressive MS was not successful. In 1983 we published that cyclophosphamide helped patients with progressive MS, but it was clear that cyclophosphamide was not as effective in patients with primary progressive MS or the later stages of secondary progressive MS. I reviewed the MRI scan and although there were changes consistent with MS, there was no active enhancement or evidence of inflammation.

"What is the strongest drug you have for MS?" she asked.

"It's our chemotherapy drugs," I said.

"I want to do everything I can to stop the disease," she said. "I still want to be able to play doubles tennis."

I went over the different chemotherapy options and we decided to treat her with cyclophosphamide because of our long experience with the drug. She had now been taking cyclophosphamide for almost three years.

Her husband escorted her, and she walked holding onto a cane and his arm for support. She walked the twenty-five-foot carpet in twenty seconds but needed bilateral support. I introduced her to Susan and we sat in the exam room. I looked through her chart and did not bring up the idea of playing tennis—this was clearly out of the question.

"How do you feel?" I asked.

"I feel okay," she said, "though I get a little tired from the treatments."

"Do you feel that the treatments are helping you at all?"

"Hard to say," she said.

"I think she just keeps getting worse," her husband said. "It's harder and harder for her to walk."

"Do you think we should continue with the chemotherapy?" she asked.

I looked at her chart and saw how she had worsened over time. There had never been much evidence of inflammation on her MRI.

"We gave it a good shot," I said, "but I think it may be better to stop the chemotherapy now."

After a long discussion we elected to give her a course of steroids and put her on one of the injectable medications plus a drug used in other diseases to help nerve cell degeneration. She was thankful that she no longer was taking chemotherapy and hopeful that the new medication might help her.

After she left Susan turned to me and said, "What a difference from the college professor who did so well with cyclophosphamide."

"MS has different stages and different types," I said. "We used to group all of the MS patients together, but they are clearly different. I think people like her present one of our greatest challenges—people with the primary progressive form of the disease or people who have entered the progressive or late progressive stages. We were a bit naive when we first obtained our results with cyclophosphamide and concluded that it helped progressive MS. It helps when there is an early inflammatory component to the disease, but doesn't work well in patients like her or in later-stage disease."

"It's not your fault," Susan said.

THE NEXT PATIENT was a man in his forties named Michael Case who carried a cane and was accompanied by his brother. The patient had a neatly trimmed beard. I greeted them in the hallway and escorted them to the examining room. Before entering the room, I turned to Susan.

"Did you notice anything unusual about them?" I asked.

"Not really," Susan said.

"They're identical twins," I said.

"Do they both have MS?" she asked.

"They do," I said, "though Michael has accumulated much more disability than his brother Mark."

We went in and I adjusted my seat to face them both in the examining room. "How are you guys doing?" I asked.

"Mark is always trying to catch up with me," Michael said, "but this is one arena in which I hope he doesn't make it," he said, pointing to his cane.

"This disease is the only way I've been able to beat you at golf," Mark said as he gave his brother a punch on the shoulder, and they both laughed.

I had been following Michael for almost ten years, and his disease was reasonably well controlled with injectable medication and occasional pulses of steroids, though he had had a number of attacks early in the course of his illness and now required the use of a cane. Recently he had complained of severe fatigue. On one of his visits he told me his twin brother had experienced some transient numbness in one of his hands but brushed it off, thinking that he had just slept wrong on his arm.

"Do you remember when I suggested that your brother bring you in for an evaluation?" I asked Mark.

"I was afraid," Mark said. "The last thing I wanted was to hear that I had MS. When you raised the possibility that I take an MRI to see if I had MS, it took me almost a year to decide."

"What finally made you decide to get the MRI?" Susan asked.

"I saw what was happening to my brother and I knew there were treatments that could help," he said. "In the end I decided that it is better to know what you're facing than to live with uncertainty."

I went over their exams and then Susan and I checked their latest MRI scans. It was clear that Michael's MRI showed more disease involvement than Mark's.

"Can we look at the MRI that allowed you to make the diagnosis of Mark's MS?" Susan asked.

We brought up the image on our computer and compared it to his latest MRI. His initial scan showed three small but typical areas of MS in the brain and one small area in the spinal cord. In the three years that I had been following him, there had been no change on the MRI. Like his twin brother, Mark was on an injectable medication.

Michael and Mark illustrated important points about nature versus nurture in MS. The study of twins has shown that genetics is important but not the whole story. For example, of one hundred people with MS who have identical twins, approximately thirty-five of the identical twins will get MS. For fraternal twins, only about five will get MS. In other words, genes are important, but they aren't the entire story. It is generally believed that MS is a complex genetic disease, and a search is under way to find genes that are linked to the disease. Family and adoption studies demonstrate that being related to someone with MS carries an elevated risk.

"Although there are family clusters of MS," I said to Susan, "most MS is sporadic."

"If you have MS and no one else in your family has MS," Susan asked, "what are the chances that one of your children will get MS?"

"There is a slightly increased risk," I said, "but not something I would worry about."

Back in the room, we finished up with the twins. Because their lives were intertwined, in a way they got two chances at the disease. Mark felt the pain of Michael's disability, and Michael was happy that by getting MS first it had allowed early diagnosis and treatment in his brother. We prescribed medication for Michael's fatigue.

"My getting MS helped you," Michael said. "Just make sure you don't waste it."

"What does that mean?" Mark said.

"Keep taking your medicine and work on your golf swing."

"I NEED A note from you," the patient said, prior to having her do the twenty-five-foot walk.

"Do the walk first and then we will talk about the note."

"How fast should I go?"

"Go as fast as you can."

I timed her on the stopwatch and she did it in 4.2 seconds down and 4.1 back.

"If we ever have an Olympic competition for walking on a twenty-five-foot red carpet, you will win the gold medal," I said. Then I went over her exam, which was completely within normal limits.

"What was it you needed the letter for?" I asked, thinking it was something related to work or a trip and she needed a letter to take syringes on the plane.

"I am climbing Mt. McKinley," she said. "There is a group of us MS patients who are all taking one of the injectable medicines and doing well, and we want to show what you can do with MS."

She had been on injectable medication for almost five years. "Have you climbed mountains before?" I asked.

"No, but I have been in training."

"Do you feel you can do it?" I asked.

"Of course," she said.

"From the MS standpoint, there is no reason you can't, and I will be happy to give you a letter," I said.

"I have a digital camera," she said as she left the exam room. "I can e-mail you pictures from the mountain."

WE SAW THREE other patients on various forms of injectable therapy—the different types of interferons and glatiramer acetate. I saw these patients once a year. Their MRIs were stable, they were working and raising their children, and they all were also participating in our natural history study in which we planned to follow one thousand MS patients over time and find out why some did well and some didn't.

There were two patients who were participating in a clinical trial of a new MS drug. We examined them and filled out the study forms. Clinical trials in MS had come a long way from my first trial with plasma exchange in 1978.

Then, as I looked at the chart of the next-to-last patient, I slowly shook my head.

"What's the problem?" Susan asked.

"The next patient I have followed for almost twenty years. Unfortunately, she has not done well, and there is little more to do for her at this time. She has some cognitive problems and she is very bitter about her MS and there is nothing I can do about it."

"In what way is she bitter?" Susan asked.

"You'll see," I said.

I brought in Margaret Williams, who was now close to seventy years old. She was in a wheelchair and had use of only her right arm. When I finished the examination, we began the same discussion we had been having for years. When we first had the discussion, I was taken aback by it, but now it was expected. Nonetheless, it was never easy.

"Do you have something to make me walk again?" she began.

"No," I said.

"Why don't you say yes?"

"If I said yes, it wouldn't be truthful."

"It would give me hope."

"But you know it wouldn't be true."

"You are not working hard enough," she said.

"I am working as hard as I can," I said. "Do you have any other things you would like to discuss?"

"Just make me all better. I'd like to run in a marathon. I bet you would too."

"We are doing research to try to help you."

"I figured that's what you are doing."

"It isn't easy, so it takes time," I said.

"You're smart. Speed it up."

"What if I told you the problem is tough and we can't solve it that easily?"

"Can we solve it at all?"

"Yes, but it will take time."

"That doesn't quite satisfy me."

"What would satisfy you?"

"Flying. I want to walk away from this chair, though I must say it is a very nice chair."

"Your mood seems good."

"It's because I am seeing you."

"How are your grandchildren?"

"I have a girl and a boy. I can't visit my daughter because she lives in Philadelphia and has a lot of steps."

"Anything else you'd like to talk about?"

"Yes, make me all better."

"All I can promise is that we are doing all that we can."

"You are not working hard enough."

"I will give you an appointment to see me in a year."

"Can't I see you earlier?"

"How often?"

"Every day."

"I promise you we will continue to work hard, and I am wishing you well."

"Is it possible to wish a little harder?"

THE LAST PATIENT was a woman on one of the injectable medications who was doing reasonably well. She had a recent attack that we treated with steroids, and she had recovered. The attacks were not as severe as before she went on the injectable, and she hadn't accumulated any disability. She had a positive attitude and participated in one of the MS support groups. When we were done she said to me, "I don't want you to feel bad that you haven't cured MS. I know you are trying as hard as you can and that alone makes me feel good."

"Quite a difference between her and Margaret Williams," Susan said to me later.

"They are both speaking from their hearts," I said. "They are not speaking to me personally; they are speaking to everyone working on MS. So," I said to her, "are you happy you're here?"

"I am not afraid of the challenge," Susan said.

IN JUNE 2005, our MS center moved to new quarters, a short distance from the Brigham and Women's Hospital. At the new MS center, we had a larger room for infusions, three 25-foot carpets to measure walking, and consolidated space for our clinical research staff. We revamped the flow of patients to create what the hospital called "the ideal patient experience," including separate check-in and check-out desks and more efficient phone coverage. In my view, the ideal patient experience for someone with MS is an easy-to-take treatment that halts the progression of the disease.

It was five years since our center had opened, and the field was clearly changing. The big event in 2005 for our center was the approval of natalizumab (Tysabri) and then its withdrawal. The opportunity to use a new drug such as natalizumab provided an important option to those patients who were not responding well to current therapy. Now that natalizumab was unavailable, one of the major focuses of our discussions and strategic planning centered on what treatments we would use in the place of natalizumab to control MS. Of course, we used chemotherapy drugs such as cyclophosphamide and mitoxantrone, but we were searching for the next generation of drugs to modulate the immune system that were less toxic and easier to administer. Two drugs that were available to us to administer because they were FDA-approved for other indications were an oral immunosuppressant called mycophenolate (Cell Cept), which was used in transplantation, and daclizimab (Zenapax), a monoclonal antibody given by infusion that was used to treat certain leukemias and had shown promising results in initial studies in MS patients. Thus, we began using these drugs in some of our patients.

In 2005, Susan Gauthier became a junior faculty member at our center, and we had a new crop of fellows who came to learn about MS. Wednesday lunch was the time for our clinical conference, during which difficult or informative cases were presented, and when we discussed the latest MS research and strategic directions for testing new MS therapies. The conference began by showing MRI films. Rohit Bakshi, a neurologist and MRI expert who was recruited to our center from Buffalo, went over the films. The first MRI showed a large mass in the front of the brain. It looked like a tumor but had features that suggested that it could also be MS. We discussed the case, and there was divided opinion in our group of whether it was MS or a tumor. We all agreed, however, that a biopsy of the mass was indicated. A biopsy had been done, and it indeed showed MS. The patient was treated with high doses of steroids and placed on one of the interferon drugs and did well. The next MRI showed multiple spots on the brain, and the patient had been sent to our center for evaluation of which treatment to begin.

"This doesn't look like MS," Bakshi said, because the location of the spots in the brain were not typical of MS. "It probably is vascular disease," he said. With the advent of MRI imaging, it has become commonplace for us to evaluate patients for whom spots are discovered on the brain that were not MS. "That's one way to cure MS," I said.

Sal Napoli was a new fellow whose decision to become a clinical MS specialist was because an immediate family member of his had the disease. He presented a case whose main difficulty was cognition. Although MS is not a disease with major cognitive problems such as Alzheimer's disease, it has become clear that cognition can be impaired in some MS patients. Maria Houtchens, another fellow, was interested in the MRI basis for cognitive problems in MS. Although MS had always been thought to only affect the white matter of the brain, there was now evidence that gray matter could be involved, and that areas of the cerebral cortex could be affected. Our center was testing drugs to help cognition, including drugs that had been used in Alzheimer's disease, such as donepezil (Aricept) and memantine (Namenda). Bonnie Glanz, a neuropsychologist, was performing cognitive testing on the MS patients in our natural history study.

"What's unique about this next case?" Bakshi asked, as he put up another set of MRIs.

"It looks like typical MS to me," Dave Dawson said.

"It is," Bakshi said, "but this MRI is from a boy who is eleven years old. Athough MS did not usually occur in children, childhood MS was becoming increasingly recognized. Tanuja Chitnis, a neurologist and immunologist in our center, established a program to follow cases of pediatric MS and she commented on this case. "The MS drugs we use for adults are also used in children," she said, "and they work." The National MS Society was in the process of setting up a network of centers to understand and treat childhood MS.

Our Wednesday noon conference ended with a discussion of clinical trials. Guy Buckle, who was director of clinical care at our MS center, reported on progress being made in a trial of an interferon drug that was given orally, called interferon tau. One of our major goals was to find orally active drugs for MS. Samia Khoury, who co-directs the MS center with me, showed promising immunology blood test results of a drug called CTLA-4 Ig that we hoped could be a drug to reset the balance of the immune system (induce tolerance) without major side effects. We grappled with which drugs to test in the progressive forms of the disease, one of the hardest to treat. Susan Gauthier had been treating some primary progressive patients with an anticancer drug called rituxumab (Rituxan) that was also being studied in a large multicentered trial.

What had changed dramatically in the five years since our MS center opened was the difficulty in performing clinical trials that included a placebo group and in finding appropriate patients to enter clinical trials. The issue of placebo groups in clinical trials was affecting the entire field, as it was hard to deny treatment to patients, and ethical issues had been raised by human study committees whether patients can indeed take a placebo when there are other drugs available. Sandy Cook, one of our research nurses, outlined the inclusion and exclusion criteria for each of the clinical trials. As it became more difficult to enroll patients in clinical trials in the United States, drug companies were looking for patients overseas.

The conference ended with my encouraging doctors to fill out their electronic forms and to obtain blood samples for the laboratory and for our natural history study. When I looked at the overflowing crowd of people at our Wednesday noon conference, I took it as a clear indication of the vitality of the efforts of the scientific community to find a cure for MS and evidence that we were indeed beginning to win the battle of taming the monster.

16

GOING FOR THE CURE

CURE.

The word *cure* is one of the most powerful words in medicine. It sits on one side of a complex equation that connects doctor to patient, science to disease, an equation that in the end is reduced to an individual suffering from illness who yearns to be relieved of that suffering, to be healed, to be cured. The idea of a cure is ingrained in every person who has ever been ill—and that's all of us. Ingrained in all our psyches is the idea that something—magically, even miraculously—will relieve us of our ailment, our suffering.

Cures are everywhere. There are biblical stories of cures, of blind people who could suddenly see, of crippled people who could suddenly walk. There are tabloid stories of cures for cancer in someone who took a mixture of herbs or specially extracted substances that were obtained in Mexico, or somewhere in Europe, or at a clinic in California. There are stories of people with a terminal illness, given only a few months to live, who are somehow alive years later, because of their will, prayers, diet, or positive attitude. Friends or relatives tell us of an acquaintance who was sick and then somehow was no longer sick. There are stories in the medical literature of dramatic healing. And then there is the story of polio, once a feared disease, now no longer the scourge it was. And the story of insulin, no need to die if you have diabetes; and the story of penicillin, truly a miraculous cure for bacterial infectious disease. Cures abound. People who were dying didn't die. People who were becoming crippled no longer became crippled. Unfortunately, cures are more difficult for chronic diseases such as multiple sclerosis.

In today's modern age we are confronted by the endless marketing and spin of advertisers trying to find ever more sophisticated ways to market their products. The same is true for many diseases and the word *cure*. There are countless campaigns to find the "cure"—the cure for diabetes, the cure for breast cancer, the cure for AIDS, the cure for Alzheimer's disease, the cure for prostate cancer, the cure for heart disease, and of course the cure for multiple sclerosis.

"We seldom cure anything," Henry Claman said to me on one of my recent visits to Colorado, where we discussed the latest in immunology and we talked about the current treatment of MS with the injectable medications. "In my mind there are only two conditions that medicine can truly cure," he said. "First, there are infections by bacteria. We are able to eradicate the bacteria with antibiotics. That's how we cure strep throat with penicillin. The other times we cure with surgery. If someone has a localized cancer, we can cut it out. The cancer is no longer there. The patient is cured."

The word *cure* is used with abandon when we talk about raising money for research, when we talk about working on the problem. "Our goal is to cure MS," the banners read. But the word *cure* is also one that is scrupulously avoided whenever new treatments are announced that help MS patients, to warn patients that the treasure we are all looking for has not yet been found.

When we announced our results with cyclophosphamide we were careful to warn that we had not found the cure. When Betaseron was approved Barry Arnason wrote that this was not the cure. When Copaxone was discussed at the FDA one of the committee members was quick to point out that although Copaxone helped MS patients, it was not the cure. The same was true when I participated in the FDA approval of the chemotherapy drug mitoxantrone for multiple sclerosis. Please make sure that everyone knows this isn't the cure.

"We cured polio and put a man on the moon," people say. "Shouldn't we be able to cure MS?" Of course, we never cured people who had polio, we brought polio under control by preventing it, by stopping it from ever happening. What, then, does one say to the countless MS patients who ask when we will find a cure? What do we say to patients who bring in a newspaper clipping or Internet article that describes the latest breakthrough?

We will find a cure for multiple sclerosis, of that I am certain. The only question is when and how it will happen.

I was once asked to speak at the New York chapter of the National Multiple Sclerosis Society. There was a general session for the MS patients and I was also asked to speak at a special brunch for major donors.

"How long do I have?" I asked.

"Twenty minutes," the woman said.

I thought for a second and then proposed the following title: "How Are We Going to Cure MS?" It was obviously a bold title. Besides, how could one even begin to address such a question in twenty minutes? Nonetheless, I wanted to explore the idea so as to put into perspective the concept of curing MS. In the end, isn't that what we all want, a cure for MS? The story of MS will not be complete until a cure is found.

I stood at the podium and began my talk by asking, "What do we mean when we talk about a cure for MS?"

A woman in the front row raised her hand and said, "A cure is when someone in a wheelchair gets out of their wheelchair and is normal again."

"Are there any other definitions?"

A man in the back said, "I think a cure is when someone comes down with MS and we are able to stop it so that it never progresses."

"Any other definitions?"

A young woman in a yellow dress said, "To me a cure is when nobody ever gets the disease."

I asked for other definitions, but there were none.

I myself could not think of other definitions. The three answers that were given covered the crucial conceptual aspects of the problem. I then answered the questions in order.

"The first type of cure," I said, "would involve rebuilding a damaged nervous system. Indeed, there is active research in this area, and we have hopes of helping those with spinal cord injury with growth factors and research in stem cells and myelin repair. We all hope that one day we will be able to help people get out of their wheelchairs and walk normally. This, of course, was the hope of Christopher Reeve. It is something that is not just around the corner, but it is possible. I have learned over the years never to say never in terms of the hope of a patient, no matter how severe the illness. Indeed, even if we think the chances are infinitesimally small that something might happen, one never knows." I cautioned that such a cure would take a long time, even though there was a great deal of research in that area. I didn't mention the point that if MS was caused by the immune system and only the nervous system was repaired, the repaired nervous system would just be attacked again by an abnormal immune system. This is indeed what happened in diabetes when pancreatic cells from a normal twin were transplanted into the twin who had diabetes. The healthy twin's pancreatic cells were only destroyed by the abnormal immune system of the twin that had diabetes.

One of the newest hopes for human disease is stem cell research. Stem cells are immature cells that have not yet differentiated and offer the potential of a limitless source of cells for research and therapy. For therapy, scientists envision injection or transplantation of stem cells to replace or repair a damaged organ. One such example is growing stem cells into insulin-producing cells (or islets) and injecting them into people who have lost their insulin-producing cells to diabetes. Islet transplantation works in diabetes, but there is a limited source of islets. Stem cells have the potential to solve the problem. I participated in a special task force on stem cell research in MS held in Washington, D.C. The task force found that there is a high likelihood that research on all types of stem cells will improve our understanding of the MS disease process and lead to new pathways for therapeutic intervention. However, we felt it is not likely that direct injection of stem cells into areas of the brain and spinal cord affected by MS will be a useful treatment for MS in the near term. Because MS affects multiple parts of the brain and spinal cord, and because the brain is such a complicated organ, it will be difficult to replace or repair damaged tissue in MS using a stem cell approach. Nonetheless, stem cells offer a wonderful opportunity for disease modeling, drug screening, and understanding basic mechanisms in MS. Because stem cell research holds great promise for people affected by MS, the MS Society recommends funding and supporting all meritorious stem cell research, including research using human embryonic stem cells.

Unfortunately, false hope is raised with reports of new therapies that "cure" MS in mice or rats. I read a recent report that showed a picture of a paralyzed mouse and a normal mouse with the caption, "A mouse paralyzed with a variant of multiple sclerosis walked again thanks to a new treatment aimed at errant immune cells." This is unlikely to happen to someone with MS who has been paralyzed for a long period of time. Nonetheless, people with rapid worsening or in the midst of an attack can recover their ability to walk again. Another report I read had the headline, "Cure for MS is at hand," even though the findings in the report dealt with the basic function of a myelin structure that was only tangentially related to MS.

"The second cure," I said, "relates to what I call a different C word, and that word is *control*." When a person comes down with MS, what we worry about are more attacks and progression of the illness. If there was only one attack, MS wouldn't be such a bad disease. With evidence that drugs that affect the immune system can reduce MS attacks by restoring the immune balance and new knowledge that we may also need drugs to combat the degenerative processes that are triggered by inflammation, it should be possible to devise drug combinations that keep the immune system in balance, block degeneration, and thus prevent the accumulation of disability. A person is not cured as the disease is still in the body, but the disease is controlled. I believe this is the fastest road to a cure, and one that I envision beginning to happen over the next decade (it may be actually happening now in some patients) as we better understand the immune basis of the disease, are able to monitor the disease better, and develop less toxic and more effective drugs that modulate the immune system and prevent degeneration. Our MS center website (www.mscenter.net) provides an up-to-date review of the next generation of drugs being tested for MS (see glossary).

I had a discussion with a friend whose brother died of MS in which we debated the concept of cure. My friend argued that if you can control the disease with medicine, it in fact was cured. Her daughter argued that the disease was not cured as long as the patient needed medicine to control it. The disease was still there, ready to strike. She believed that one had to devise a treatment that changed the immune system and excised the monster so that medication was no longer needed. Only then would the disease be truly cured.

I had a similar discussion with Samia Khoury, a neurologist and immunologist who directs the clinical immunology laboratory and experimental treatment programs at our MS center. We have worked together since 1990, and our discussions are not theoretical, as we must choose which new drugs to try in MS. We are planning to test new combination therapies, oral agents, drugs that affect degeneration, and drugs to induce immunologic tolerance that do not have the potentially toxic side effects of chemotherapy. We both believe that in the future the first treatment MS patients will receive is a strong immune modulator that induces immune tolerance followed by long-term maintenance therapy. We need more physicians like Samia, trained in both clinical medicine and laboratory investigation, to carry out the translational research needed to go from lab bench to bedside.

"The third cure," I said, "is the ultimate cure. If MS is truly a disease of the immune system and it begins by shaping the immune system in childhood in people with a certain genetic background and with a certain environmental exposure, we might be able to eradicate MS. We would understand who is susceptible, and we would treat them before they got the disease—we would vaccinate them like we vaccinate against polio, we would stop the monster from being created." I emphasized that this was a very long experiment that would take a generation. First we would have to completely understand the immune and genetic basis of the disease and then vaccinate children and wait twenty-five to fifty years to prove that the vaccination worked and they never got MS. Henry Claman said to me, "You can effectively treat a disease without knowing its cause, but you need to know its cause to prevent it." It was much easier to prove the polio vaccine worked, because one only had to vaccinate children in the fall, and show that in the summer they didn't contract polio. It was a one-year experiment. The ultimate MS experiment is going to be a longer one, like testing a medication designed to prolong life of people that are 60 to determine if it allows them to live to 120. One has to wait sixty years to find out the results of the experiment.

Steve Hauser and I have begun outlining a plan to lay the groundwork for the third cure. It is based on the fact that certain populations are at increased risk for MS. The idea is to begin studying these at-risk populations before they get MS to understand the factors that cause it and identify markers that predict who is at the highest risk to get MS. If we are successful, the next step is to treat an at-risk population and prevent them from getting MS. John Richert, a neurologist and MS specialist, is the new director of research and clinical programs at the National MS Society; he follows the lineage of Harry Weaver, Byron Waksman, and Stephen Reingold. He would like to set up such a program and initiate long-term studies of factors associated with MS, as was done in the Framingham heart study.

Five years later, I received another invitation to speak at the New York chapter of the National MS Society. A lot had changed in five years. This time I spoke about curing MS to the audience of MS patients and about the ethics of doing clinical trials to a smaller group of major donors. A problem confronting the MS community was how to study new drugs in MS in the face of FDA-approved drugs.

A NUMBER OF years ago at an MRI meeting in Oxford, England, I began asking scientists and clinicians studying MS the "MS genie" question. It goes like this. "Imagine you are on the beach, find a bottle, rub it, and a genie appears. You discover that it isn't an ordinary genie that gives you three wishes. It is the MS genie, and you get only one wish. You can have the answer to any question about MS that you pose. Think about it—what is the most important question about MS for which we don't know the answer or that you personally are burning to know?"

People usually laugh when they hear the question, but then they become serious. I like the exercise because it forces people to think. Implicit in the exercise, of course, is confronting the issue of what research experiments are needed to answer the question we pose to the genie. We must be our own MS genies. Should we be putting special effort into a particular series of experiments? Is the question we are asking answerable at this time, or is more technology needed? I think the genie exercise is something all scientists should undertake about the problem they are investigating.

What answers did I get? Is there a top ten list of "Answers we'd like from the MS-genie," just like on the David Letterman show? Actually, after collating all the answers, I came up with a top-five list. I have ranked them below from bottom to top, though in many ways they are interrelated.

5. *Viruses*

Some wanted to know once and for all whether there is one MS virus, especially scientists with a virology background. As we know, most scientists have given up looking for one MS virus, even though most believe viruses trigger T cells and MS attacks. Knowing that there is no MS virus would not change current MS research or our path to a cure, but finding an MS virus would have a major impact in our approach to the disease. I do not believe that there is an MS virus to be found, though absence of proof is not proof of absence.

4. *Progression*

Many wanted to know what happens in MS patients when the disease becomes progressive. Related to that question was this one: if we treat aggressively and shut down all the inflammation at the first sign of the disease, will it stop later progression and degeneration? This is the central question of current MS therapy, and the initial answers are encouraging. People treated early with anti-inflammatory drugs do appear to have less progression later on. It is a long experiment, though, lasting ten to twenty years. One must treat aggressively at the first sign of the disease and demonstrate that patients do not enter the progressive stage ten to twenty years later.

3. *Benign MS*

Why do some people have benign disease and others have malignant disease? It appears that MS is not one disease but variants of a disease process. It is known that for some, MS is not a debilitating disease. On occasion MS is discovered at autopsy in a person who had no symptoms during his or her lifetime. If we knew the factors that were missing or present that affected severity of disease, it would give clues to therapy and prognosis.

2. Immune dysregulation?

Many people, of course, turned the question around and asked me what my MS genie question was. Being an immunologist, I want to ask the genie what is wrong with the immune system in MS. Is there a defect in the induction of T cells that regulate the immune response? Is there an abnormal signal that lets disease-inducing cells expand? If we knew what was wrong with the immune system, we would know how to devise better drugs, and immune abnormalities could be corrected at the first sign of disease to prevent progression. Knowing what was wrong with the immune system would also give clues for prevention, maybe even an MS vaccine.

1. Initial trigger

The most popular question concerned the very first event that starts the disease. Was it a virus that triggered a T cell to go into the brain? Was it something in the brain itself that became abnormal and led to the T cell attack? Did MS start in the brain or outside the brain? When I asked how we might answer the question, most agreed that it was difficult, since by the time we see a patient with MS, the disease has already started. The current theory is that it is a T cell that enters the brain and begins the process, though there is no formal proof.

In a way the MS genie question is analogous to getting a group of scientists together to create a position paper on our strategies to find a cure for the disease. This was done in 1972, when the National Advisory Commission on Multiple Sclerosis was established by Public Law 92–563, signed by the president on October 25, 1972. The commission's report was issued on February 11, 1974. Over twenty-five years later, in 1998, the National Multiple Sclerosis Society commissioned the Institute of Medicine in Washington, D.C., under the aegis of the National Academy of Sciences, to carry out a similar task. Their report, entitled "Multiple Sclerosis: Current Status and Strategies for the Future," was published in 2001.

What did the reports say, and how different were they after twenty-five years? How do they relate to the "MS Genie" question? Based on the reports, what type of effort is needed to cure MS and how should we allocate our time, energy, and resources? Are we on the right track?

Looking through the reports, what is remarkable is how much progress has been made over the past thirty years, how many of the lines of research have borne fruit, and how the paths being investigated have not changed dramatically. What has driven us closer to a cure over the past three decades has been a better understanding of the basic biology underlying the immune system and the brain, the development of crucial tools such as MRI imaging to study the brain, and the involvement of the pharmaceutical industry. If anything has changed dramatically, it is the technology now available to study MS.

The specific scientific recommendations of the two committees will not be detailed here, but both involve the detailed study of the immune system in MS patients and animal models; a better understanding of the pathology of MS, of how nerve cells are damaged and ways in which drugs to protect and repair the nervous system can be developed; and the application of a burgeoning new technology that allows one to screen for the presence of thousands of genes and proteins on a single chip.

Both committees recognized the importance of recruiting young investigators into the field and the sharing and dissemination of information about research. The 1974 committee suggested creating special abstracts to disseminate information about MS research, which was undertaken for many years in a now obscure publication called *MS Indicative Abstracts* and which, of course, is now provided by the technology of the electronic age. If one queries the National Library of Medicine under "multiple sclerosis," 31,000 citations appear. Searching for a virus or infectious agent was part of both reports, even though the search to date has been negative. And even though MRI had not yet developed in 1974, the earlier committee emphasized the importance of developing means to image the nervous system in MS, a breakthrough that occurred independent of MS-directed research and won a Nobel Prize in 2003 for discoveries that were made in the 1970s and led to MRI imaging.

What can we expect in the next decade? First, many clinical trials of new drugs that are directed at modulating the immune system are in progress or planned. Hopefully we should have drugs available to us that are stronger and do not have the side effects of chemotherapy drugs. The injectable medications that are now taken by MS patients are likely to be replaced by oral drugs. Many trials of combination therapy, like that done in cancer, will be performed. And although the marketing techniques and often narrow scientific focus of the pharmaceutical industry can be frustrating at times to the MS researcher, the industry is currently investing hundreds of millions of dollars to develop and test new therapies for MS. This is a recent development and one that is crucial for FDA approval of new drugs for MS. I believe the efficiency of our free-market society will hasten MS drug development. Furthermore, a relatively rapid route to develop drugs for MS is to test drugs that have been approved for other conditions that may benefit MS patients. In this instance, the long development process is not needed. Indeed, many of the drugs currently being tested in MS are already approved for another indication. Over one hundred fifty clinical trials are currently in progress, most focused on modulating the immune system according to the latest listing by the National Multiple Sclerosis Society.

The path forward is relatively clear regarding immune therapy of MS. The next therapeutic frontier is the development of treatments that protect nerve fibers from damage and of treatments that can help rebuild myelin. Clinical trials are in their infancy in this area but are actively being pursued.

Advances in technology often drive discovery. The newest technology being applied to MS is broadly called genomics, research in which one screens thousands of genes or proteins to determine which ones are associated with MS. There is no specific hypothesis. One does not ask whether a particular chemical or protein is elevated or reduced in MS, one simply measures thousands of them that are embedded on chips and that react with blood samples or brain samples from MS patients. It is a giant fishing expedition made possible by technology. The results are so complex that they require special computer programs for analysis. Nonetheless, positive results are being uncovered. Initial results suggest certain protein patterns linked to different types of MS and MS treatments. In addition, certain gene and protein patterns are being observed in the brains of patients with MS.

When I began my research career, the concept of "big science" had not yet evolved. That has changed. One of the recommendations of the 2001 Institute of Medicine report was to initiate expensive, large-scale projects, and some are under way. With the deciphering of the human genome, we now have the possibility of understanding the complex genetics of MS. A consortium of genetic researchers in the United States and Europe is collecting thousands of DNA samples from MS patients and their families to apply the latest genetic mapping techniques in search of genes that are linked to the disease. In Munich, the Sylvia Lawry Center, named for the woman who founded the National Multiple Sclerosis Society in 1946, is collecting a massive database of clinical and MRI data from MS clinical trials. The MS lesion project based at the Mayo Clinic is collecting biopsy specimens from around the world to better understand the pathology and subtypes of MS. Large databases such as the European Database for Multiple Sclerosis (EDMUS), based in Lyon, are providing new insights into MS. In addition, clinical trials have entered the realm of big science both conceptually and financially. They require international cooperation, multiple investigator sites, and at times in excess of a thousand subjects. A recently initiated trial of combining interferon with glatiramer acetate, for example, will cost the National Institute of Neurological Diseases and Stroke in excess of $30 million, and corporate trials of this magnitude are not uncommon. Joyce Nelson, the new CEO of the National MS Society, announced five-year, multimillion-dollar grants to four centers to support integrated research on strategies to protect and repair damaged nervous tissue in MS. The awards were an outgrowth of a special workshop that brought scientists together to strategize on how to deal with the degenerative component of the MS process.

Where do the money and organizational structure to cure MS come from? In the United States, the basic and clinical research enterprise is fueled by the National Institutes of Health and the National Multiple Sclerosis Society, which provide over $110 million per year for MS research. Outside the United States, a network of thirty-four national MS societies provide support for research as do numerous private foundations. Pharmaceutical companies now provide large sums of money for clinical trials and laboratory research. Finally, one cannot ignore the billions of dollars of support for basic biomedical research provided by the National Institutes of Health, without which there would be no basic findings to apply to diseases such as MS.

Someone once asked me what I would do to hasten the day that we cured MS. The purpose of this book is not to become involved in the politics of research funding. But if I had one suggestion, it would be to facilitate the development of MS clinical and research centers such as ours and then provide means for centers to collaborate and pool clinical, MRI, immunological, and genetic data on MS patients. MS centers with areas of expertise that receive long-term support can make a major contribution. One such example is the support of MRI research at Queen Square in London by the MS Society of Great Britain.

How will we know that we are indeed curing MS? The biggest challenge is the amount of time it takes to test a therapy and then to know that it is truly making a difference. There are many natural history studies of MS prior to the current era of therapy. We now need such studies in the era of treatment to determine if the course of MS has been altered. We have initiated such a study at our MS center and will be tracking one thousand patients over ten years, measuring clinical attacks and progression and doing MRI imaging and blood studies on a yearly basis.

Our ultimate goal is to treat a person who comes down with MS and show that twenty years later that person is still normal. To help speed this, we will study patients with MS in five-year blocks, like five-year survival rates for cancer. What will greatly help us in showing that we are curing MS is to develop surrogate markers that we can measure early in treatment that will link to the effectiveness of treatment over the long term. Such markers may be blood tests and MRI patterns. It will be like knowing that if blood pressure is controlled, there will be fewer heart attacks. After the polio vaccine was developed, there was not as great a need for iron lungs. In the future, we will measure our progress in MS by the decrease in the number of wheelchairs purchased by those diagnosed with MS.

In actuality, I think there are people now who have had their MS cured according to the second definition. These are patients who have responded to currently used injectable medications and to chemotherapy; for ten years their disease has been controlled. I sometimes think we should construct a red thermometer like those used to reach the goal in fund-raising to show what percentage of MS patients are stable five years, ten years, and fifteen years on treatment.

There are those who feel we are on the wrong track—that MS is not even an autoimmune disease, or that the current focus of research is wrong. I don't agree, but if we are on the wrong track, there must be an alternative hypothesis and experiments to test it. And until we are curing MS to a greater degree than we are now, there will always be those who say we are on the wrong track, and there will be quack cures and crazy theories like tooth amalgam, sucrose- and tobacco-free diets, and intravenous injections of cells from yeasts.

What does the MS patient do while we are in the process of curing MS? Living with MS can be very frustrating. Much of the research and breakthroughs reported by the press do not translate immediately into treatment for those suffering from the disease. There is not an easy answer. Nonetheless, there is hope, as many new treatments are being tested for MS. In addition, therapy to help MS symptoms and rehabilitation programs can make a big difference in the life of an MS patient, as can patient support networks.

It was a clear day in Colorado, with the sun brighter than it ever is in Boston. I was driving with my mother to visit Normie again. I had seen him approximately a year before and visited him just after my talk in New York. He knew I was coming and was there waiting for me on the porch of the facility as I drove up. He spotted my car just as I spotted him sitting in his wheelchair.

He manipulated his electric scooter as we entered the home. I told him that it looked like he was losing some hair, and he retorted, "God gave some people brains and other people hair." We both laughed. As we went down the corridor he introduced me to people by saying, "This is my friend who is so smart and knows everything." As we settled into his room I brought him up to date on our research, and he told me I looked great on the *Montel Williams Show.* Two other people joined us, a woman who worked in the facility and a man who was also in a wheelchair and who I learned had the primary progressive form of MS.

Normie continued with his jokes and his jovial attitude. "Did you hear about the doctor who gave his patient six months to live, and when he found out the patient couldn't pay the bill, he gave him another six months?" After spending a few minutes telling me about problems with his bladder and discussing the medications he was taking, he then remarked that someone asked him whether he smokes after sex, and he answered that he never looked. We all laughed again. Next he brought me up to date on his family. His daughter with his first wife was now an attorney and had two children. His first wife was now remarried. Another one of his daughters was pregnant. His sister was living in California.

"What about a cure, Big Skinny?" he said to me finally. Both Normie and his friend John leaned forward waiting for my answer, trying to catch a glimmer of hope that might help them.

I told them about my talk in New York and the different definitions of cure and how we are moving closer and closer. I thought of what there is that we can realistically do that might help them, but the options are limited. I thought of our understanding of the disease and wondered whether in my lifetime I would see the day that a treatment is given so that MS doesn't occur anymore. I realize that the experiment is too long for me to see it in my lifetime.

As I looked at Normie, I thought back on my life—all the research I have done, all the patients I have treated. I think of the experiments planned in the laboratory and the new clinical trials we are planning. I think of our new MS center, with state-of-the-art facilities. When I started thirty years earlier as a neurology resident looking after John Saccone, what had I imagined for myself? What had I hoped to achieve, and what actually happened? I've poured the best years of my life into trying to understand and treat MS. I've become an MS doctor with the resources to investigate the disease at the highest level. I've made contributions, I've influenced the field, but as I looked back I came to realize how difficult the biology is, how small a piece I have been in the big picture, how progress in understanding the disease has been driven by forces I cannot control—basic biological discovery, technology such as the MRI scan. I have taken solace from those people that I helped and felt pain for all those I couldn't. I have been bothered by the pettiness of scientists and their egos, the marketing tactics of drug companies, the politics of funding. I have felt the frustration of patients trying to understand and cope with a disease their doctors often don't understand and have trouble explaining. I have taken joy in being able to participate, to make advances, to train scientists to continue in a field that has broken open. The monster is truly being tamed. I have seen it with my own eyes. The problems are better defined; the technology has advanced. People are clearly being helped. It makes no difference to me who finds the answers, though I know that the chase and the games and the prizes are needed incentives for progress.

As I looked at Normie, I thought back to my office at the Center for Neurologic Diseases in Boston, where I have a view over the skyline of downtown Boston and could almost see Massachusetts General Hospital and the tiny laboratory where I had performed my first EAE experiment in Barry Arnason's lab over thirty years ago and drew blood from my first MS patient. Over the years my laboratory grew and moved from one research building to another in the Harvard Medical area that straddles Longwood Avenue. The Longwood Medical area itself would grow exponentially with countless new research buildings and hospital expansions. Some call it the Wall Street of medicine.

I have hung on my walls six sayings over those thirty years that have helped provide inspiration and guidance. There is Churchill's advice to never, ever quit, and John Enders' quote about how scientific discovery is usually made step by step, based on what had gone previously not by one moment of epiphany.

For my fellows I have a quote from Francis Crick when he received the Nobel Prize for discovering the structure of DNA. I put it on the wall to encourage open and critical dialogue in the lab. "Politeness," Crick said, "is the poison of all good collaboration in science. If one figure is too much senior to others, the serpent politeness creeps in. A good scientist values criticism almost more highly than friendship; no, in science criticism is the height and measure of friendship. The collaborator points out the obvious with due impatience. He stops the nonsense."

For myself, as my career grew, and for my fellows as they rose to the rank of professor, I had a quote from a former neurology professor of mine: "The early days in the laboratory are the golden years for many scientists. Later, the volume of mail, telephone calls, number of visitors, organizational activities, including committees by the dozen, and demands for lectures, reviews and community activities grow insidiously and will destroy the creativity of the scientist if unopposed."

I had just come across a quote from Louis Pasteur that I was planning to place on the wall. I liked the idea of a quote from Louis Pasteur, as my laboratories were in a research building on Avenue Louis Pasteur across from the Harvard Medical School. It had special meaning to someone like myself that was performing basic research in the laboratory but was interested in disease: "To those who devote their lives to science, nothing can give more happiness than increasing the number of discoveries, but one's cup of joy is full when the results of one's studies immediately find practical applications."

Then there are two things that I myself wrote. The first deals with the focus that is needed to solve a scientific problem, the focus needed to one day cure MS: "Only with discipline and a singularity of purpose can one achieve anything. Otherwise, one is bound to drift aimlessly, pushed by the tides to some distant shore; or worse, trapped by a fate that circles endlessly in the same place."

The last saying on the wall I wrote only recently, a reflection on life and on coping with illness: "It is a given that life is not perfect and one of the major challenges for every person is to live one's life well given the imperfections they must face."

I realized that Normie would likely never walk normally again, and even though I will have spent my entire life working on MS, I won't get to see the ultimate cure.

"I can't promise that we will get you out of your wheelchair in this lifetime," I told him. "Although there is always a chance," I said, never wanting to take away hope completely.

"But let's make a pact," I told Normie. "We have no firsthand knowledge that there will be a next life, but let's assume that there is and that in that life we will meet again. By then we will truly have the cure for MS."

"Sounds good to me," Normie said.

"I will get to know how the disease is cured, and you will be out of your wheelchair," I said with excitement in my voice. "We'll eat banana splits, ride together on your motor scooter, and chase girls."

We all laughed and there were tears in everybody's eyes.

"It is great to see you, Big Skinny," Normie said.

"It is great to see you, Little Fats," I answered.

Glossary

ACTH Adrenocorticotropic hormone, a hormone that stimulates the adrenal gland to make cortisone. Previously used to treat attacks of MS.

Adjuvant A compound used to boost the immune response. It is mixed with myelin proteins (MBP, MOG, PLP) and injected into animals to induce EAE, the animal model of MS.

Antibody Part of the immune system that binds specifically to its target and helps fight infection. A polio vaccine induces antibodies that bind to and inactivate poliovirus, but not smallpox virus.

Axon Nerve fibers in the brain and spinal cord that carry electricity.

B cell A white blood cell that makes antibodies.

Brainstem A brain structure that controls eye movements and facial muscles.

Bystander suppression Suppression of inflammation by factors released by T cells.

Cerebellum A brain structure that controls coordination and balance.

CNS Central nervous system; includes the brain and spinal cord.

CSF Cerebrospinal fluid.

Cyclophosphamide Generic name for Cytoxan, a chemotherapy drug used to treat worsening MS.

Double-blind A clinical trial in which neither the patient nor the doctor is aware of which treatment is being given.

EAE Experimental allergic (or autoimmune) encephalomyelitis, an animal model of MS.

Gadolinium A substance injected into the vein at the time of an MRI. If there is brain inflammation, it leaks into the brain and lights up on the MRI (gadolinium enhancement).

Gamma globulin A type of antibody that is increased in the spinal fluid of MS.

Glatiramer acetate Generic name for Copaxone, an FDA-approved drug for MS.

Immunosuppression Drugs that suppress or dampen the immune system.

Interferon A natural substance that interferes with replication of viruses and also affects the immune system. Interferon beta 1a (Avonex, Rebif) and Interferon beta 1b (Betaseron) are FDA-approved drugs for MS.

Lesion A damaged area in the brain.

Lumbar puncture A spinal tap in which fluid is removed from the spinal canal.

Lymphocyte A type of white blood cell that is part of the immune system.

Macrophage A white blood cell that has the property of engulfing particles, a scavenger cell. It causes damage in MS by stripping myelin from the myelin sheath.

MBP Myelin basic protein, a protein of the myelin sheath used to induce EAE.

MHC Major histocompatibility complex, a structure needed to trigger T cells.

Mitoxantrone Generic name for Novantrone, an FDA-approved chemotherapy drug for worsening MS.

MOG Myelin oligodendrocyte glycoprotein, a protein of the myelin sheath used to induce EAE.

MONOCLONAL ANTIBODY An antibody made by B cells that have been immortalized so they produce unlimited amounts of a very specific antibody in culture vessels.

MS ATTACK Symptoms of nervous system dysfunction that usually come on over a week and resolve in two to six weeks.

MYELIN A fatty protein substance that covers the nerve fibers in the nervous system and is damaged in MS.

NATALIZUMAB Generic name for Tysabri, a monoclonal antibody given by monthly infusion that has been shown to reduce relapses and prevent disability in MS.

OLIGOCLONAL BANDS Pattern of elevated gamma globulin in the spinal fluid that is often seen in MS.

OLIGODENDROCYTE A cell in the nervous system that makes myelin and is damaged in MS.

OPTIC NEURITIS Inflammation of the optic nerve that causes blindness, a common first symptom of MS.

PLACEBO A sham pill or treatment.

PLP Proteolipid protein, a protein of the myelin sheath used to induce EAE.

PRIMARY PROGRESSIVE MS A type of MS that involves slow worsening from the onset.

RELAPSING-REMITTING MS A pattern of MS of attacks followed by recovery.

SALINE Salt water.

SECONDARY PROGRESSIVE MS A type of MS that involves slow worsening. It is preceded by the relapsing-remitting pattern.

SERUM The liquid part of the blood.

T CELL A white blood cell that is part of the immune system and is the major regulator of the immune system. It is named for originating in the thymus. There are many types of T cells. It is believed that T cells that attack the myelin sheath initiate MS.

TH1 CELL A T cell that makes the chemical interferon gamma. It causes EAE and is believed to cause inflammation in MS.

TH2 CELL A T cell that makes the chemical IL-4. It has anti-inflammatory properties and may regulate Th1 cells.

TH3 CELL A T cell that makes the chemical TGF-beta. It has strong anti-inflammatory properties and may regulate both Th1 and Th2 cells.

THYMUS A gland under the neck where T cells develop.

VENTRICLES Cavities in the brain where spinal fluid is made.

WHITE BLOOD CELLS Cells in the blood that fight infection.

For further information on MS, the reader is referred to the National Multiple Sclerosis Society (www.nationalmssociety.org), the Multiple Sclerosis International Federation (www.msif.org), and the Partners Multiple Sclerosis Center at Brigham and Women's Hospital (www.mscenter.net).

BASIC AND CLINICAL RESEARCH in MS is moving at a rapid pace. If the reader is interested in the cutting edge of what is happening in MS before it appears in the next edition of *Curing MS,* please visit our MS center website (www.mscenter.net). At the end of each year, I write a review of progress under the heading "What's new in MS research and clinical trials."

Acknowledgments

A large number of physicians and scientists have made significant contributions to our understanding and treatment of MS. Those that have been referred to in this book are those that I have worked with or come in contact with, and those whose research related to the aspects of the MS story chronicled in this book. There are countless others who are not mentioned in this book who are crucial parts of the MS story, and whose contributions have made this book possible.

MS research requires a great deal of money, and I would like to acknowledge the NIH and National Multiple Sclerosis Society without whose support my work and the work of all those in the field could not be done. Furthermore, there are countless individuals whose generous philanthropy has enabled us to break new ground in our research. A personal thanks goes to each one of them for their support. Special thanks to the Nancy Davis Foundation, which has supported our MS research for more than a decade, and to the Montel Williams MS Foundation.

My closest collaborators in the study of MS over the years are David Hafler, Samia Khoury, and Vijay Kuchroo. They are wonderful scientific colleagues and have become family. They not only critically reviewed the manuscript, but critically review everything I do. As one of the quotes hanging on my wall says, they "stop the nonsense."

The doctors, nurses, and staff at the Partners MS Center provide an environment that gives the best care to our patients while at the same time facilitating our research. Special thanks to Lynn Stazzone and Sandra Cook.

The scientists and postdoctoral research fellows in the laboratory provide the advances in basic and applied research that are translated into new treatments and that provide the raw material needed to ultimately cure the disease. Special thanks to Ruth Maron.

I would like to thank those who gave of their time to be interviewed: Ruth Arnon, Steve Hauser, Ken Johnson, Roy Lobb, Steve Marcus, Hill Panitch, Mike Panzara, Don Paty, Irit Pinchasi, Rick Rudick, Al Sandrock, Michael Sela, Nancy Simonian, Dvorah Teitelbaum, and Byron Waksman.

Those who provided critical comments on the manuscript: Robert Bishop, Cynthia Campbell, Joanne Chertok, Allison Cohen, Alastair Compston, Dave Dawson, Marika Hohol, Janet Kaplan, Joe and Rhoda Kaplan, Larry Levitt, Joe and Evy Megerman, Jennifer Robinson, and Byron Waksman. Special thanks to Stephen Reingold.

Lane Zachary, my literary agent, provided support and guidance and nurtured the book from an idea across a bumpy road to a completed project. Having her has made all the difference. Emily Loose, my former editor at Crown, provided clear direction that was critical in shaping the tone of the book, and Rachel Kahan at Crown provided crucial support in the last stages of the project. Leslie Epstein, head of the creative writing program at Boston University, has supported my writing from its inception, and every year has asked me when I was going to write this book.

My sons, Dan and Ron, gave insightful critique and support that only sons can give to a father. My wife, Mira, who is now part of our MS center, has been with me for the entire journey, a journey I could not have made without her.

Finally, I thank the patients for their courage and for teaching me so much about life.

Index

A

B

C

D

E

I

J

K

About the Author

Howard L. Weiner, M.D., is the Robert L. Kroc Professor of Neurology at Harvard Medical School. He is founder and director of the Partners Multiple Sclerosis Center at the Brigham and Women's Hospital in Boston, Massachusetts, and codirector of the Center for Neurologic Diseases at the Brigham and Women's Hospital. Dr. Weiner has been on the forefront of developing new therapies for MS for more than twenty-five years. He has published widely in the scientific literature on multiple sclerosis and is the author of *The Children's Ward,* a novel, and *Neurology for the House Officer,* a medical text now in its seventh edition. He lives with his wife, Mira, in Brookline, Massachusetts.